Calculus and the Personal Computer

C. H. Edwards, Jr.

The University of Georgia

Prentice-Hall, *Englewood Cliffs, New Jersey*   07632

Library of Congress Cataloging-in-Publication Data

Edwards, C. H. (Charles Henry), (date)
   Calculus and the personal computer.

   Bibliography: p.
   Includes index.
   1. Calculus—Data processing. 2. Microcomputers.
I. Title.
QA303.E2235 1986   515'.028'5536   85-19123
ISBN 0-13-112319-X

Editorial/production supervision: *Maria McColligan*
Designed by: *Walter A. Behnke, Judy Winthrop, Jayne Conte*
Cover design: *Bruce Kenselaar*
Cover photo: *David Coons*
Manufacturing buyer: *John B. Hall*

Printed in the United States of America

10 9 8 7 6 5 4 3 2 1

ISBN 0-13-112319-X   01

Prentice-Hall International (UK) Limited, *London*
Prentice-Hall of Australia Pty. Limited, *Sydney*
Prentice-Hall Canada Inc., *Toronto*
Prentice-Hall Hispanoamericana, S.A., *Mexico*
Prentice-Hall of India Private Limited, *New Delhi*
Prentice-Hall of Japan, Inc., *Tokyo*
Prentice-Hall of Southeast Asia Pte. Ltd., *Singapore*
Editora Prentice-Hall do Brasil, Ltda., *Rio de Janeiro*
Whitehall Books Limited, *Wellington, New Zealand*

Contents

Preface

I wrote this book to encourage the use of the personal computer for both the introduction and the illumination of basic concepts of introductory calculus. One of the main appeals of calculus is its power to solve interesting real-world problems, but elementary students often find the numerical computations in such problems a barrier rather than an aid to understanding. The computational power of the computer now makes many exciting applications of calculus more widely accessible to such students. Computer techniques that rely largely on high school mathematics can be used to solve problems that formerly required a prior knowledge of calculus. These problems can, in turn, be used to motivate the introduction of the concepts of calculus itself. This approach employs the personal computer to introduce the central concepts of calculus in a "see it on the screen" manner that many students will find more tangible and concrete than conventional textbook expositions. The book is intended to serve either

1. As a computer supplement to a standard calculus course (or as the text for a computer laboratory associated with such a course);
2. As the text for an experimental course on calculus with computing; or
3. As a source for self-study by people who would like to understand better both calculus and their personal computers.

The formal prerequisite to reading this book is a reasonable knowledge of high school (precalculus) mathematics. No calculus at all appears (explicitly) in Chapter 1, and the remaining chapters are logically self-contained with respect to the concepts of calculus—it being our goal to employ the computer to motivate and introduce these concepts, rather than merely to illustrate them. However, I occasionally refer for further discussion to Edwards and Penney, *Calculus* (C. H. Edwards, Jr. and David E. Penney, *Calculus and Analytic Geometry*, second edition, Prentice-Hall, 1986), though the analogous sections of any other standard calculus textbook can be used instead.

RANGE OF TOPICS

Together, the five chapters span the computational highlights of a typical single-variable calculus course. Chapter 1 discusses several methods of solving equations (together with applications of these methods), winding up with "zoomsolve" computer graphics. This material introduces concretely the idea of iteration and the concept of convergence in the specific case of a sequence of approximations to a solution. In this pragmatic context the reader assimilates both elementary programming skills and a practical acquaintance with limits of sequences.

Limits of sequences are treated more generally in the initial section of Chapter 2. This leads naturally to limits of functions and to derivatives, both discussed computationally—that is, in terms of how we write and use programs to actually compute their values. Chapter 2 concludes with computer graphics solutions of applied maximum-minimum problems. Chapter 3 introduces the integral as the area under a curve. In a computational approach we find it quite natural to proceed directly from the definition of the integral to numerical integration techniques such as the trapezoidal and Simpson approximations, rather than to the exact techniques and applications that typically are discussed next in a standard calculus course. Chapter 4 is devoted to the numerical summation of infinite series, principally for the purpose of accurately calculating values of the familiar elementary functions. In regard to exploiting the computer as a pedagogical tool, the whole calculus course may offer no better opportunity than the chapter on infinite series, traditionally the most abstract one in a calculus text. My own teaching experience indicates that numerical summation is the perfect antidote for the student whose initial impression is that one infinite series is about the same as any other one.

In the first three sections of Chapter 5 we see that the computer enables us to apply quite elementary ideas of velocity and acceleration—with little calculus explicitly needed—to solve an impressive range of problems involving skydivers, batted baseballs and pitched curves, artillery projectiles, missiles, and rockets. Finally, in the last section of Chapter 5 we observe that similar ideas provide a beginning for the subject of numerical solution of differential equations.

This brief collection of topics summarizes the changes in the standard introductory calculus course that seem most likely to result from the spreading use of microcomputers for instructional purposes—more discussion of numerical solution of equations to illustrate iterative processes; limits of sequences preceding limits of functions; more emphasis on numerical integration; a more computational approach to infinite series; and earlier numerical treatment of elementary differential equations.

HARDWARE AND SOFTWARE

My own experience with the computer has taught me that no flowchart or outlined algorithm has the convincing tangibility of a program that actually runs. The illustrative programs in this book are written in IBM-PC BASIC. A diskette containing all the programs listed herein is available to instructors who use an IBM Personal Computer or compatible. However, these programs are readily converted to versions of BASIC used with other microcomputers. Indeed, with the exception of the graphics programs in Sections 1.6 and 2.5, they can even be rewritten for the inexpensive pocket computers that now are widely available.

Although a specific choice of programming language was necessary, the emphasis throughout is on the personal mastery of computational skills, rather than on the use of a prepared package of programs. Therefore the most instructive use of the book might be with equipment that requires the students to rewrite all the programs! With an appropriate class, an enticing possibility would be the translation of our BASIC programs into nicely structured Pascal. At the opposite extreme, the inclusion of screen displays illustrating the output of all the listed programs should make it possible to read this book for self-study without even using a computer.

APPROACH AND LEVEL

I assume the modest acquaintance with BASIC (though no actual programming experience) that is fairly standard among college-bound high school graduates with scientific interests. The goal is to develop programming skills hand-in-hand with mathematical skills, taking advantage of each to enhance the learning of the other. A principal feature of this book is the devotion of considerable effort to writing programs that are attractive and instructive to read, and therefore can play a center-piece role in the exposition. The format of each program is designed to spotlight its underlying mathematical idea (without regard to whether the program structure is top-down, bottom-up, or sidewise). Whenever a choice presents itself, I favor mathematical clarity over programming elegance or efficiency. Technical questions regarding accuracy and efficiency are deferred to the appropriate course in numerical analysis, so as not to dilute our concentration on the fundamental concepts of calculus. However, we check the accuracy of our programs in situations where known results are available for comparison—thus the many approximations to e and π—and attempt to include examples that develop an intuitive feeling for the accuracy of the computational methods we introduce.

Above all, perhaps, I hope to communicate a comfortable familiarity with the personal computer as a handy everyday tool for mathematical problem-solving. The highly structured and heavily documented programs that computer scientists generally advocate have their place in general-purpose programming for a batch-processing, compiler-oriented mainframe environment. But the interactive and interpretive world of the personal computer calls for brief and uncluttered programs that are readily altered in a cut-and-fit manner to serve the purpose at hand. The typical exercise in the problem sections calls for the execution of a program that is listed in the text, either as it is or with an appropriate modification. Each of these listed programs (as printed) has been run successfully on an IBM Personal Computer, and all program listings and printouts were photocopied directly so as to eliminate the possibility of typing errors.

Though only a single name appears on the title page, this book is really the end product of a wholesome and rewarding family project. Alice F. Edwards not only did the word-processing but also actively shared the labor (and joy) of composition and ferreted out many rough spots that otherwise would have survived in the final manuscript. John K. Edwards is the family expert on the inner workings of the personal computer and, in the case of a number of programs in the book, he must be thanked for the fact that they run correctly.

C.H.E

Solution of Equations

1

1.1 INTRODUCTION

The solution of equations is a staple topic in elementary mathematics. One learns in school to solve simple linear equations, pairs of linear equations in two unknowns, and the general quadratic equation

$$ax^2 + bx + c = 0, \tag{1}$$

which has two solutions (possibly equal or complex) given by the *quadratic formula*

$$x = \frac{-b \pm \sqrt{b^2 - 4ac}}{2a}. \tag{2}$$

Solving equations of various types remains a topic of importance in mathematics at every level. However, the equations we encounter in real-world applications of mathematics are seldom as simple as those mentioned above. As an illustration, the following example describes a concrete physical situation that is governed by a *cubic* equation—one of degree *three*, in contrast to the quadratic equation (1) of degree two.

Example

The radius of a large cork ball (Fig. 1.1) is 1 ft, and its density is one-fourth that of water. Determine the depth x to which this ball sinks when it floats in water.

1

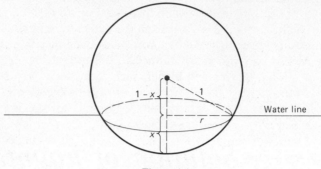

Figure 1.1

Solution. According to Archimedes' *law of buoyancy*, a floating body displaces an amount of water whose weight precisely equals that of the body. Since water is four times as dense as our cork ball, the ball will float with exactly one-fourth of its total volume of $4\pi(1)^3/3$ ft^3 submerged beneath the waterline. Hence, our problem is to determine the depth x indicated in Figure 1.1, so that the volume of the submerged portion of the ball is $V = (1/4)(4\pi/3) = \pi/3$ ft^3.

As indicated in the figure, the plane of the water surface intersects the spherical ball in a circle of radius r. The submerged portion cut from the sphere by this plane is a "spherical segment" with *height x* and *radius r*. The volume V of such a spherical segment is given by the known formula

$$V = \frac{\pi x}{6}(3r^2 + x^2), \tag{3}$$

where $\pi \approx 3.14159$, as usual (later in the book you will learn *why* this *is* the value of π rounded to five decimal places).

If we set $V = \pi/3$ in (3) we would have two unknowns, r and x, so we must first eliminate one of them. To do this, we use the right triangle shown in Figure 1.1. Its hypotenuse is the radius 1 of the ball, and its legs are r and $1 - x$. Hence the Pythagorean theorem says that

$$r^2 + (1 - x)^2 = 1,$$

so

$$r^2 = 1 - (1 - x)^2 = 2x - x^2.$$

When we substitute this value of r^2 in (3), and also equate V to $\pi/3$, we get

$$\frac{\pi}{3} = \frac{\pi x}{6}[3(2x - x^2) + x^2],$$
$$2 = x(6x - 2x^2),$$
$$x^3 - 3x^2 + 1 = 0. \tag{4}$$

So the depth x must satisfy this *cubic* equation.

Formulas for the solution of third- and fourth-degree equations—in the same spirit as the quadratic formula, in that they involve roots of combinations of the coefficients—were discovered in the sixteenth century. However, these formulas are too complicated for practical use, and therefore are rarely seen today. Moreover, it was proved in the early nineteenth century that for equations of the fifth degree and higher, such formulas *do not exist.*

How then, as a practical matter, do we go about solving a cubic equation like (4)? To begin, we can observe that the physical situation implies that $0 < x < 1$. Certainly, x must be positive, and $x < 1$ because less than half the sphere of radius 1 is submerged. In addition, we can show mathematically that Equation (4) must have a solution between 0 and 1. This is a consequence of the following important theorem that we will have frequent occasion to apply.

Intermediate Value Theorem.   Let $f(x)$ be a continuous function defined on the closed interval $a \leqq x \leqq b$. If the values $f(a)$ and $f(b)$ at the two endpoints have opposite signs, then $f(c) = 0$ for some number c between a and b.

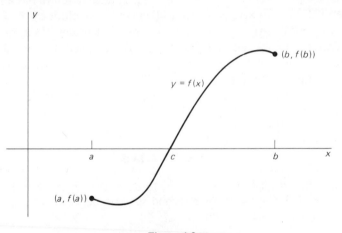

Figure 1.2

We will not dwell here on the meaning of the word "continuous" (see Section 2.4 of Edwards and Penney, *Calculus*), except to remark that every polynomial function such as $f(x) = x^3 - 3x^2 + 1$ is continuous. Therefore, the intermediate value theorem applies to polynomial functions. Figure 1.2 illustrates the case in which $f(a) < 0$ and $f(b) > 0$ [the other case being $f(a) > 0$ and $f(b) < 0$]. The curve $y = f(x)$ starts at the point $(a, f(a))$ below the x-axis, and ends at the point $(b, f(b))$ above the x-axis. The theorem says simply that it must cross the x-axis at some point c between a and b, so $f(c) = 0$ at this point.

The closed interval $a \leq x \leq b$ on the x-axis is often denoted by $[a, b]$, and we say that $f(x)$ *changes sign* on $[a, b]$ if the values $f(a)$ and $f(b)$ have opposite signs. With this notation and terminology we can restate the intermediate value theorem as follows: *If the continuous function $f(x)$ changes sign on $[a, b]$, then the equation $f(x) = 0$ has a solution in the open interval $a < x < b$.* This open interval is commonly denoted by (a, b).

To apply this theorem to our *cork ball equation*,

$$f(x) = x^3 - 3x^2 + 1 = 0, \tag{5}$$

we merely calculate the values $f(0) = 1 > 0$ and $f(1) = -1 < 0$. Thus $f(x)$ changes sign on $[0, 1]$, and therefore has a solution in $(0, 1)$. We can go a bit further by calculating

$$f\left(\frac{1}{2}\right) = \left(\frac{1}{2}\right)^3 - 3\left(\frac{1}{2}\right)^2 + 1 = \frac{3}{8} > 0.$$

Now we see that $f(x)$ changes sign on $[1/2, 1]$ and therefore has a solution between $1/2$ and 1. In Section 1.2 we will find that this solution is $x \approx 0.6527$. Thus our cork ball sinks in water to a depth just under $2/3$ ft, or 8 in.

Throughout this chapter we will use the cork ball equation as a "test case" for a number of different techniques for solving equations. The development of these techniques will, in turn, illustrate the diverse ways in which the power of the personal computer can be employed both to illuminate and to solve problems in mathematical analysis.

PROBLEMS

1. Use the intermediate value theorem to show that the cork ball equation $f(x) = x^3 - 3x^2 + 1 = 0$ has a second solution between -1 and 0, and a third solution between 2 and 3.

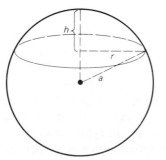

Figure 1.3

2. Figure 1.3 shows a spherical segment of height h and radius r cut from a sphere of radius a by a plane at distance $x = a - h$ from the sphere's center. Use Formula (3) to show that the volume of this segment is

$$V = \frac{\pi h^2}{3}(3a - h).$$

3. An interesting problem of Archimedes was that of using a plane to cut a sphere into two segments with volumes having a given (preassigned) ratio. Suppose that a plane at distance x from the center of the unit sphere $(a = 1)$ cuts it into two segments, one having *twice* the volume of the other. Then use the formula of Problem 2 to show that x satisfies the cubic equation

$$3x^3 - 9x + 2 = 0.$$

4. Use the intermediate value theorem to show that the cubic equation of Problem 3 has three real solutions: one in $(-2, -1)$, one in $(1, 2)$, and the third one (the one we are looking for) between 0 and 1.

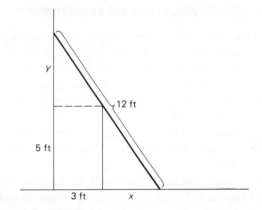

Figure 1.4

5. A 5-ft fence stands 3 ft from a tall wall. A 12-ft ladder leans across the fence and just touches the wall (see Fig. 1.4). If x denotes the distance of the foot of the ladder from the base of the fence, show that x satisfies the quartic (fourth-degree) equation

$$x^4 + 6x^3 - 110x^2 + 150x + 225 = 0.$$

Hint: Note that the two smaller triangles in Figure 1.4 are similar, and apply the Pythagorean theorem to the large right triangle.

6. Show that the quartic equation in Problem 5 has one solution between 2 and 3, and another between 6 and 7. It also has two negative solutions, but the two positive solutions give the *two* possible positions of the ladder.

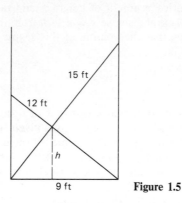

9 ft   **Figure 1.5**

7. Figure 1.5 shows a 12-ft ladder and a 15-ft ladder leaning against the opposite walls of an alley 9 ft wide. What is the height h of their crossing point above the ground? *Hint:* Three parallel lines cut proportional segments from any two transversal lines.

1.2 EVALUATION AND TABULATION OF FUNCTIONS

Tables of values of functions are familiar to anyone who had been exposed to logarithms or trigonometric functions. By the fifteenth century developments in astrology(!), astronomy, and navigation had progressed to the point that very complete and extensive trigonometric tables were needed to compute accurately the positions of stars and planets in the sky. During this period numerous mathematicians devoted essentially their whole working lives to the compilation of such tables. Consequently, there arose the idea of constructing calculating machines that could ease this tedious and time-consuming labor.

One of the first such machines was built by Leibniz (co-inventor with Newton of the calculus in the late seventeenth century), who said: "It is unworthy of excellent men to lose hours like slaves in the labor of calculation." But only in our own time have personal computers come to be widely available to do this "work of slaves."

One of the simplest applications of computers is the tabulation of functions—calculation of the values of a given function $f(x)$ for a specified range of values of x. But even so simple a computational facility is surprisingly useful. In this section we describe the application of tabulation techniques to solve some rather formidable equations. This approach relies on the number-crunching power of the computer rather than on sophisticated mathematical concepts.

To illustrate this brute-force approach to the solution of equations, we consider the cork ball equation

$$f(x) = x^3 - 3x^2 + 1 = 0. \qquad (1)$$

x	f (x)
-5	-199
-4	-111
-3	-53
-2	-19
-1	-3
0	1
1	-1
2	-3
3	1
4	17
5	51

Figure 1.6 Values of $f(x) = x^3 - 3x^2 + 1$

Figure 1.6 shows a table of values of $f(x) = x^3 - 3x^2 + 1$ for integral values of x from $x = -5$ to $x = 5$. Scanning the right-hand column of this table, we see that $f(x)$ changes sign on the intervals $[-1, 0]$, $[0, 1]$, and $[2, 3]$. Therefore, the intermediate value theorem tells us that Equation (1) has a solution in each of these three intervals. Any cubic equation has three solutions, although one pair of them may be either equal real roots or conjugate complex roots, and in special cases the three solutions of a cubic equation can coincide so that it actually has only a single distinct (real) solution. It follows that Equation (1) has precisely three real solutions, and we have located them in the three intervals above.

The solution between 0 and 1 gives the depth to which the cork ball of Section 1.1 sinks when it floats in water. But *where* in the interval $(0, 1)$ does this root lie? To focus our attention on this interval, we run the program

```
10  PRINT "x", "f(x)"
20  PRINT
30  FOR I = 0 TO 10
40      X = I/10
50      F = X^3 - 3*X*X + 1
60      PRINT X, F
70  NEXT I
80  END
```

For $I = 0, 1, 2, \ldots, 10$, the variable $X = I/10$ takes the values $0, 0.1, 0.2, \ldots, 1$, and line 50 calculates the corresponding values of $f(x)$. The output of this simple program is shown in Figure. 1.7. Observe that $f(x)$ changes sign on the interval $[0.6, 0.7]$. Hence we now know that our solution lies somewhere between $x = 0.6$ and $x = 0.7$.

We could narrow our focus still further by changing line 40 of the program to

$$40 \qquad X = 0.6 + I/100$$

so that it will calculate and display the values of $f(x)$ for $x = 0.60, 0.61,$

x	f(x)
0	1
.1	.9709999
.2	.888
.3	.757
.4	.584
.5	.375
.6	.136
.7	-.1269999
.8	-.4080001
.9	-.701
1	-1

Figure 1.7 Values of $f(x) = x^3 - 3x^2 + 1$

$0.62, \ldots, 0.70$. However, we may later want to investigate the solutions of Equation (1) in the intervals $(-1, 0)$ and $(2, 3)$. Instead of repeatedly revising the program above, it obviously would be more efficient to write—once and for all—a general tabulation program that we can use to calculate and display the values of $f(x)$ at equally spaced points in *any* desired interval $[a, b]$.

```
100 REM--Program TABULATE
105 REM--Tabulates the values of f(x) on the
110 REM--interval [a,b] by increments of h =
115 REM--(b - a)/n.  The function f(x) must be
120 REM--edited into line 250.
130 REM
140 REM--Initialization:
150      INPUT "ENDPOINTS A,B"; A,B
160      INPUT "NUMBER OF SUBINTERVALS"; N
170      LET H = (B-A)/N
180      PRINT "x", "f(x)"
190      PRINT
200      LET X = A
210 REM
220 REM--Loop to calculate values:
230 REM
240      FOR I=0 TO N
250         F = X^3 - 3*X^2 + 1
260         PRINT X, F
270         X = X + H
280      NEXT I
290 REM
300      END
```

Listing 1.8 Program TABULATE

Program TABULATE is shown in Listing 1.8. The programs listed in this manner were written on an IBM Personal Computer. The first line of each such program contains (in capital letters) the name under which it may be saved on a work disk. If this has been done, the BASIC command LOAD "TABULATE" will load the program in Listing 1.8 into your computer's memory.

You can build your own program library for this book by saving each listed program on your own diskette. If you simply wish to enter and run selected programs, you may delete the REM (remark) statements that document the programs. Most of the programs in this book are written so that they can be entered (ignoring REM statements and formatting) on any BASIC computer (including pocket computers), perhaps with trivial modifications such as the replacement of semicolons with commas.

Lines 150 and 160 of Program TABULATE call for you to input the endpoints of the interval $[a, b]$ and the desired number n of subintervals. The FOR-NEXT loop in lines 240–280 calculates and displays the values of x and $f(x)$ for $x = a, a + h, a + 2h, \ldots, a + nh = b$. If we want to produce hardcopy results (rather than monitor displays) we replace the PRINT commands in lines 180, 190, and 260 with LPRINT.

Figure 1.9 shows three successive runs of Program TABULATE, each with $n = 10$ subintervals. The first run, with $a = 0.6$ and $b = 0.7$, shows that $f(x)$ changes sign on $[0.65, 0.66]$, so the solution of $x^3 - 3x^2 + 1 = 0$ we are looking for is between $x = 0.65$ and $x = 0.66$. Hence we choose $a = 0.65$ and $b = 0.66$ for the second run, and find that the solution is between $x = 0.652$ and $x = 0.653$. The third run shows that it is between 0.6527 and 0.6528, but (by interpolation) much closer to 0.6527. Thus the solution of the cork ball equation in the interval $[0, 1]$ is $x = 0.6527$, accurate to four decimal places.

Note on formatting: As written in Listing 1.8, Program TABULATE produces results like those shown in Figure 1.7, with variable numbers of decimal places in different entries. When the number of decimal places that will be best for visual inspection of the results of a particular run cannot be predicted, we write a general-purpose program without concern for the number of decimal places displayed. After a preliminary run, we then edit the PRINT command(s) as appropriate. For instance, for the second run shown in Figure 1.9, we changed line 260 to

260 PRINT USING "#.###   + #.#####"; X, F

For the third run we then replaced #.### with #.####. In Problems 1 and 2 we ask you to find similarly the other two solutions of Equation (1).

To tabulate values of a given function $f(x)$, we load Program TABULATE and then edit the formula for $f(x)$ into line 250. For instance, suppose that we want to solve the equation

$$2x^3 = 50x - 27. \tag{2}$$

We first rewrite it in the form

$$f(x) = 2x^3 - 50x + 27 = 0 \tag{3}$$

```
RUN
ENDPOINTS A,B? 0.6, 0.7
NUMBER OF SUBINTERVALS? 10

    x                   f(x)

   0.60              +0.13600
   0.61              +0.11068
   0.62              +0.08513
   0.63              +0.05935
   0.64              +0.03334
   0.65              +0.00713
   0.66              -0.01930
   0.67              -0.04594
   0.68              -0.07277
   0.69              -0.09979
   0.70              -0.12700


RUN
ENDPOINTS A,B? 0.65, 0.66
NUMBER OF SUBINTERVALS? 10

    x                   f(x)

   0.650             +0.00713
   0.651             +0.00449
   0.652             +0.00186
   0.653             -0.00078
   0.654             -0.00342
   0.655             -0.00606
   0.656             -0.00871
   0.657             -0.01135
   0.658             -0.01400
   0.659             -0.01665
   0.660             -0.01930


RUN
ENDPOINTS A,B? 0.652, 0.653
NUMBER OF SUBINTERVALS? 10

    x                   f(x)

   0.6520            +0.00186
   0.6521            +0.00159
   0.6522            +0.00133
   0.6523            +0.00106
   0.6524            +0.00080
   0.6525            +0.00054
   0.6526            +0.00027
   0.6527            +0.00001
   0.6528            -0.00025
   0.6529            -0.00052
   0.6530            -0.00078
```

Figure 1.9 Successive runs of Program TABULATE

and then edit line 250 to read

$$250 \qquad F = 2*X^3 - 50*X + 27$$

To investigate the location of the solutions, we first run Program TABU-

x	f (x)	x	f (x)
0	+27	-10	-1473
1	-21	-9	-981
2	-57	-8	-597
3	-69	-7	-309
4	-45	-6	-105
5	+27	-5	+27
6	+159	-4	+99
7	+363	-3	+123
8	+651	-2	+111
9	+1035	-1	+75
10	+1527	0	+27

Figure 1.10(a)   Values of
$f(x) = 2x^3 - 50x + 27$

Figure 1.10(b)   Values of
$f(x) = 2x^3 - 50x + 27$

LATE with $a = 0, b = 10, n = 10$, with the results shown in Figure 1.10(a). Scanning the second column for changes of sign, we see that there is one solution between $x = 0$ and $x = 1$, and another between $x = 4$ and $x = 5$. To look for the third solution of the cubic equation (3), we note that $f(0) = 27 > 0$, whereas $f(x) = 2x^3 - 50x + 27$ is negative when x is negative but sufficiently large numerically (because the leading term predominates). Hence the third solution must be negative. The results of a second run shown in Figure 1.10(b) indicate that this negative solution lies between $x = -6$ and $x = -5$. Now that we know that the three solutions of Equation (3) lie in the intervals $[-6, -5], [0, 1]$, and $[4, 5]$, we can find them to several decimal places of accuracy by successive tabulation of values of $f(x) = 2x^3 - 50x + 27$.

Thus we can use the TABULATE program to solve equations, and not merely the quadratic equations of high school algebra. Given the equation $f(x) = 0$, we first find an interval $[a, b]$ on which the function $f(x)$ changes sign, preferably with a and b being consecutive integers. Then we run Program TABULATE successively to narrow it down to smaller and smaller intervals on which the function changes sign, and within which the root therefore lies. This is the "brute-force" method of solving equations; we may call it the *method of repeated tabulation*. It yields progressively more accurate decimal *approximations* to the actual root—with a fairly general equation, we can reasonably hope only to approximate a root accurately, not to find it precisely.

The remaining sections of this chapter deal with more sophisticated methods of solution of equations. But there are occasions when brute-force tabulation is quicker and more convenient than fancier methods.

PROBLEMS

1. Use Program TABULATE to find that the other two solutions of Equation (1) are -0.5321 and 2.8794 (approximately).

2. The number $\sqrt{2}$ is the positive solution of the equation $f(x) = x^2 - 2 = 0$. Use Program TABULATE to discover that $\sqrt{2} \approx 1.414$.

3. The cube root of 5 is the solution of the equation $f(x) = x^3 - 5 = 0$ between 1 and 2. Find it by repeated tabulation, accurate to three decimal places.

4. The equation $x^3 - 4x - 1 = 0$ has only one real solution. Find it by repeated tabulation, accurate to three decimal places.

5. The equation $x^3 - 4x + 1 = 0$ has three distinct solutions. First locate the intervals in which they occur, and then find each of them (accurate to three decimal places) by repeated tabulation.

7. Find by repeated tabulation the three real solutions (accurate to three decimal places) of the cubic equation $3x^3 - 9x + 2 = 0$ of Problem 3 in Section 1.1.

8. Find by repeated tabulation the four real solutions (accurate to three decimal places) of the quartic equation

$$x^4 + 6x^3 - 110x^2 + 150x + 225 = 0$$

of Problem 5 in Section 1.1.

9. Write a BASIC program that, like TABULATE, starts out with input values of the endpoints a and b and the number n of subintervals. However, instead of printing the values of x and $f(x)$ for each of the $n + 1$ points of subdivision, its FOR-NEXT loop prints the values of x, $x + h$, and the product $f(x) * f(x + h)$ for each of the n subintervals. Write the program with a particular function such as $f(x) = x^3 - 3x^2 + 1$ (as in the text). If a and b are chosen so that $f(x)$ switches sign on $[a, b]$, you can then glance down the column of products in the output to spot the subinterval(s) where $f(x)$ switches sign. It will be the one (if there is only one) for which the product $f(x) * f(x + h)$ is negative.

10. In order to find a solution of the equation

$$f(x) = x^4 + x^3 - 1 = 0,$$

write a brief program that first asks you to input a guess x, next prints the values of x and $f(x)$, then asks you for another guess, and so on. Note first that obviously $f(0) = -1$ and $f(1) = +1$. With each guess you should attempt to interpolate (mentally) between the values of the function at the previous two guesses. Without too many guesses you should be able to find the solution accurate to four decimal places.

1.3 BISECTION AND INTERPOLATION

Now we want to "automate" the method of repeated tabulation for solving an equation $f(x) = 0$, starting with an interval $[a, b]$ on which we have observed that the continuous function $f(x)$ changes sign. By this we mean that instead of our having to inspect the tabulated output visually, the computer will determine on which subinterval of $[a, b]$ the value of $f(x)$ changes sign. It will then subdivide *this* subinterval and repeat the process, continuing until it winds up with a subinterval that (necessarily) contains a solution *and* has length less than our error tolerance for the solution. Finally, it will print out the solution.

It is most convenient to subdivide each interval into just $n = 2$ subintervals at each stage. In this case, the technique is called the *method of bisection*. We bisect the interval $[a, b]$, choose the half-interval on which $f(x)$ changes sign, bisect this half-interval, and so on.

In more detail, suppose we know in advance that $f(x)$ changes sign on $[a, b]$, and denote by $m = (a + b)/2$ the midpoint of $[a, b]$. If $f(m) = 0$, then $x = m$ is our solution, and we are finished. Otherwise, $f(m)$ has the same sign as either $f(a)$ or $f(b)$. If $f(a)f(m) < 0$, then $[a, m]$ is the subinterval on which $f(x)$ changes sign, whereas if $f(a)f(m) > 0$, then $[m, b]$ is the subinterval on which $f(x)$ changes sign.

At this point we rename as $[a, b]$ whichever of the two subintervals $[a, m]$ and $[m, b]$ is the one where $f(x)$ changes sign. The new subinterval still contains a solution of the equation $f(x) = 0$, and has *half* the length of the original interval. We now bisect the new $[a, b]$ and repeat the process.

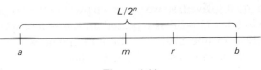

Figure 1.11

At each step the situation is as indicated in Figure 1.11—the new interval $[a, b]$ contains both its midpoint m and a solution r of the equation. After n steps the original interval—of length L, say—has been bisected n times, so the new $[a, b]$ has length $L/2^n$. Inspection of Figure 1.11 therefore makes it clear that

$$|m - r| \leq \frac{L}{2^{n+1}}. \tag{1}$$

For instance, if $L/2^{n+1} \leq 0.00001$, then m and r agree to four decimal places, so the solution of our equation equals $m = (a + b)/2$, accurate to four decimal places.

Program BISECT, shown in Listing 1.12, implements this repeated bisection procedure to solve the cork ball equation,

$$f(x) = x^3 - 3x^2 + 1 = 0.$$

```
100 REM--Program BISECT
105 REM--Solves the equation f(x)=0 by the method
110 REM--of bisection.  The function f(x) must
115 REM--be edited into line 150.  The accuracy
120 REM--desired must be specified in line 280.
130 REM
140 REM--Initialization:
150      DEF FNF(X) = X^3 - 3*X*X + 1
160      INPUT "ENDPOINTS A,B"; A,B
170      IF FNF(A)*FNF(B) > O THEN
             PRINT "TRY AGAIN!" : GOTO 160
180      PRINT "A", "B"
190      PRINT
200 REM
210 REM--Repeated bisection loop:
220 REM
230      PRINT A, B
240      M = (A+B)/2
250      IF FNF(M) = O THEN GOTO 280
260      IF FNF(A)*FNF(M) < O THEN LET B = M
             ELSE LET A = M
270      IF B - A > .00001 THEN GOTO 230
280      PRINT : PRINT "SOLUTION = "; (A+B)/2
290 REM
300      END
```

Listing 1.12   Program BISECT

The function $f(x)$ appears in the function definition of line 150, and can be changed to solve other equations. Before each bisection the current endpoints a and b are displayed, so we can keep track of the progress made. The bisection loop in lines 230–280 is repeated until $b - a \leq 0.00001$ (the error tolerance specified in line 270). The final midpoint m and the solution r therefore differ by at most $(b - a)/2 \leq 0.000005$.

```
RUN
ENDPOINTS A,B? 1, 2
TRY AGAIN!
ENDPOINTS A,B? O, 1
 A                B

 O                1
 .5               1
 .5               .75
 .625             .75
 .625             .6875
 .625             .65625
 .640625          .65625
 .6484375         .65625
 .6523438         .65625
 .6523438         .6542969
 .6523438         .6533203
 .6523438         .6528321
 .6525879         .6528321
 .6525879         .65271
 .652649          .65271
 .6526795         .65271
 .6526947         .65271

SOLUTION =   .6527061
```

Figure 1.13   A run of program BISECT

Figure 1.13 shows the output of a run of Program BISECT with $a = 0$, $b = 1$. The original interval $[0, 1]$ has been bisected 16 times, with the result that the solution is

$$x = 0.652706 \pm 0.000005,$$

so $x = 0.6527$ accurate to four decimal places.

Interpolation

Now we discuss a method that is very similar to the method of bisection, except that at each stage the new subinterval is chosen by interpolation rather than by bisection. As usual, suppose that the continuous function $f(x)$ changes sign on $[a, b]$, so its graph looks like Figure 1.14. The point x_1 is the intersection with the x-axis of the line joining the endpoints $(a, f(a))$ and $(b, f(b))$ of the curve. Thus x_1 is obtained by *interpolation* between the values $f(a)$ and $f(b)$ of the function. As indicated in the figure, x_1 will often be closer than the midpoint m to the actual solution r of the equation $f(x) = 0$.

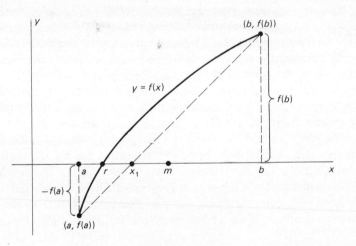

Figure 1.14

Because corresponding sides of the two similar right triangles in Figure 1.14 have equal ratios, we see that

$$\frac{x_1 - a}{-f(a)} = \frac{b - x_1}{f(b)}.$$

This equation is readily solved for

$$x_1 = \frac{af(b) - bf(a)}{f(b) - f(a)}.$$ (3)

The *method of false position* proceeds as follows. Starting with an interval $[a, b]$ on which $f(x)$ changes sign, the interpolation point x_1 is calculated using Formula (3). Unless it happens that $f(x_1) = 0$—so the "false position" of the solution is its actual position—$f(x)$ changes sign on either $[a, x_1]$ or $[x_1, b]$. Whichever one of the two subintervals this is, we choose it as the new $[a, b]$, interpolate again, and get a second interpolation point x_2. Continuing with this process, we get a sequence $x_1, x_2, x_3, \ldots$ of approximations to an actual solution r.

Whereas the method of bisection is guaranteed to "work" in the sense that preassigned accuracy can be achieved in a number of steps that can be predicted in advance using (1), the method of false position does not enjoy this advantage. The reason is that whereas the subintervals in the method of bisection approach zero in length, we have no control over the lengths of the subintervals that appear in the method of false position; their lengths need not approach zero.

If we reach the point that successive approximations are approximately equal, $x_n \approx x_{n+1}$, there is no reason to continue. If $f(x_n) \approx 0$, in addition, we have found our (approximate) solution. Although the method of false position is less predictable, it is often faster—requiring fewer steps for equivalent accuracy—than the method of bisection. This illustrates a recurring theme in the comparison of numerical approximation techniques—methods that always work are generally slow, while those that are the fastest when they work are not always successful in producing any solution at all.

Program FALSEPOS, shown in Listing 1.15, implements the method of false position. At each step the new interpolation point x, denoted by the variable XNEW, is calculated in line 240 using Formula (3). The new interval $[a, b]$ is chosen in line 260 just as in Program BISECT. In line 280 the new value XNEW is compared with the previous value XOLD to determine whether to continue. If so, XNEW becomes XOLD (line 290) for the next step.

Figure 1.16 shows the output of a run of Program FALSEPOS to solve the cork ball equation in the interval $[0, 1]$. Looking at the columns of values of x and $f(x)$, we see that $x \approx 0.6527$ is a solution, obtained in six steps as compared with the 16 steps that the method of bisection required for the same accuracy.

However, you should not jump to the conclusion that false position is always superior to bisection. When applied to the equation

$$f(x) = x^{10} - 1 = 0$$ (4)

```
100 REM--Program FALSEPOS
105 REM--Solves the equation f(x)=0 by the method
110 REM--of false position.  The function f(x)
115 REM--must be edited into line 150.  The desired
120 REM--accuracy must be specified in line 280.
130 REM
140 REM--Initialization:
150      DEF FNF(X) = X^3 - 3*X*X + 1
160      INPUT "ENDPOINTS A,B"; A,B
170      IF FNF(A)*FNF(B) > O THEN
             PRINT "TRY AGAIN!" : GOTO 160
180      PRINT "N", "    X", "  f(X)"
190      PRINT
200      N = 1 : XOLD = A
210 REM
220 REM--Repeated interpolation loop:
230 REM
240      XNEW = (A*FNF(B) - B*FNF(A))/(FNF(B) - FNF(A))
250      IF FNF(XNEW) = O THEN GOTO 270
260      IF FNF(A)*FNF(XNEW) < O THEN
             LET B = XNEW ELSE LET A = XNEW
270      PRINT N, XNEW, FNF(XNEW)
280      IF ABS(XNEW - XOLD) < .00001 THEN STOP
290      N = N + 1 :  XOLD = XNEW
300      GOTO 240
310 REM
320      END
```

Listing 1.15   Program FALSEPOS

```
RUN
ENDPOINTS A,B? 0, 1
  N                 X              f(X)

  1               .5             .375
  2               .6363636       4.282499E-02
  3               .6512969       3.709316E-03
  4               .6525855       3.116727E-04
  5               .6526938       2.592802E-05
  6               .6527028       2.384186E-06
```

Figure 1.16   A run of Program FALSEPOS

with starting interval $[0.5, 1.5]$, the method of bisection produces the obvious solution $x = 1$ in a single step, but the method of false position requires 110 steps to reach $x = 0.9998988$!

The Secant Method

Another commonly used interpolation technique is the *secant method*. The endpoints $a = x_{-1}$ and $b = x_0$ of the starting interval are used as two initial guesses for a solution of $f(x) = 0$. At each step, the next approximation x_{n+1} is defined by interpolation between (or extrapolation from) the values $f(x_{n-1})$ and $f(x_n)$ at the previous two approximations. That is, x_{n+1} is the intersection with the x-axis of the "secant line" through the points

$(x_{n-1}, f(x_{n-1}))$ and $(x_n, f(x_n))$ on the curve $y = f(x)$. By essentially the same derivation as that of Equation (3), we find that

$$x_{n+1} = \frac{x_{n-1}f(x_n) - x_n f(x_{n-1})}{f(x_n) - f(x_{n-1})}. \tag{5}$$

If we write $A = x_{n-1}$, $B = x_n$, and $XNEW = x_{n+1}$, then (5) takes the form

$$XNEW = \frac{Af(B) - Bf(A)}{f(B) - f(A)}. \tag{6}$$

Thus the secant method amounts to taking x_{n-1} and x_n as the endpoints of the new interval $[a, b]$ at each step, without regard to whether $f(x)$ changes sign on $[a, b]$.

```
100 REM--Program SECANT
105 REM--Solves the equation f(x)=0 by the secant
110 REM--method.  The function f(x) must be edited
115 REM--into line 150.  The accuracy desired must
120 REM--be specified in line 260.
130 REM
140 REM--Initialization:
150     DEF FNF(X) = X^3 - 3*X*X + 1
160     INPUT "ENDPOINTS A,B"; A,B
170     PRINT "N", "   X", "  f(X)"
180     PRINT
190     N = 1
200 REM
210 REM--Repeated interpolation loop:
220 REM
230     XNEW = (A*FNF(B) - B*FNF(A))/(FNF(B) - FNF(A))
240     PRINT N, XNEW, FNF(XNEW)
250     IF FNF(XNEW) = 0 THEN STOP
260     IF ABS(XNEW - B) < .00001 THEN STOP
270     N = N + 1 :  A = B : B = XNEW
280     GOTO 230
290 REM
300     END
```

Listing 1.17   Program SECANT

Listing 1.17 shows Program SECANT, which implements the secant method. Line 230 corresponds to Equation (6). Line 260 compares $XNEW = x_{n+1}$ and $B = x_n$, and line 270 chooses these two values as the endpoints of the interval for the next step.

Figure 1.18 shows the results of two runs of Program SECANT with the cork ball equation (2). The first run, with starting interval $[0, 1]$, gave the solution $x = 0.6527036$, which is accurate to all seven decimal places, in only five steps. The second run, with starting interval $[1, 2]$ not containing this solution, gave the same result. Comparing Figures 1.16 and 1.18, we see that in this situation the secant method is superior to the method of false position.

```
RUN
ENDPOINTS A,B? 0, 1
   N                    X                    f(X)

   1                   .5                   .375
   2                   .6363636             4.282499E-02
   3                   .653944             -3.273964E-03
   4                   .6526955             2.145767E-05
   5                   .6527036             1.192093E-07



RUN
ENDPOINTS A,B? 1, 2
   N                    X                    f(X)

   1                   .5                   .375
   2                   .6666667            -3.703702E-02
   3                   .6516855             2.685189E-03
   4                   .6526982             1.442432E-05
   5                   .6527036             0
```

Figure 1.18   Two runs of Program SECANT

However, if you run Program SECANT with Equation (4) and the starting interval $[0.5, 1.5]$, you will find that after five steps it stops at $x = 0.5343430$, which is nowhere close to a solution! It is a fact of life in the numerical solution of equations that while a particular method may work well with one equation, there will always exist another equation that requires some alternative method. This is why we need to stock our equation-solving arsenal with a variety of methods.

PROBLEMS

In each of Problems 1–5, find (accurate to four decimal places) the solution of the given equation in the indicated interval, using (a) the method of bisection, (b) the method of false position, and (c) the secant method.

1. $x^2 - 2 = 0$; $[1, 2]$
2. $x^{10} - 1000 = 0$; $[1, 2]$
3. $x^3 - 3x^2 + 1 = 0$; $[2, 3]$
4. $3x^3 - 9x + 2 = 0$; $[0, 1]$
5. $x^4 + 6x^3 - 110x^2 + 150x + 225 = 0$; $[6, 7]$
6. Run Program FALSEPOS for the equation $x^{10} - 1 = 0$ with starting interval $[0.5, 1.5]$.
7. Run Program SECANT for the equation $x^{10} - 1 = 0$ with starting interval $[0.5, 1.5]$. Can you explain what goes wrong? Note that the curve $y = x^{10} - 1$ is very flat on most of the interval $0 < x < 1$, but is very steep for $x > 1$.

8. Run Programs BISECT and FALSEPOS for the equation

$$f(x) = \frac{1}{10x - 1} = 0$$

with starting interval $[0, 1]$. BISECT will give $x \approx 0.1$ and FALSEPOS will give $x \approx 0.2$, but neither of these numbers is a solution of the given equation. What is wrong here?

9. The secant method is most effective when the initial guesses are very close to the solution. Write a program that first uses the method of bisection to produce a new interval $[a, b]$ containing the solution with $|a - b| < 0.01$, and then applies the secant method (with $x_{-1} = a$ and $x_0 = b$) to compute the solution accurate to five decimal places.

1.4 ITERATION AND REPEATED SUBSTITUTION

The equation $x^2 - 2 - 0$ certainly looks like an easy one to solve—its positive solution is simply $x = \sqrt{2}$. But what, precisely, *is* the square root of 2? The square root function in BASIC is denoted by SQR, and the command PRINT SQR(2) yields 1.414214.

However, we recall the ancient Greek proof that the number $\sqrt{2}$ is *irrational*—it cannot be expressed as a quotient p/q of two integers p and q. Hence the number $1.414214 = 1,414,214/1,000,000$ cannot exactly equal $\sqrt{2}$; it is merely the six-decimal-place approximation to $\sqrt{2}$.

But where do such decimal-place approximations come from, anyway? You may recall vaguely from grade school a certain quite tedious and boring method of calculating square roots. But the ancient Babylonians had a much better way over 2000 years ago. Here is their method: To calculate the square root $\sqrt{A}$ of a given positive number A, we start with a first estimate x_1 of $\sqrt{A}$, perhaps just the first crude guess we can make easily. For instance, for $\sqrt{17}$ we might guess $x_1 = 4$, since $(4)^2 = 16$. If we are lucky enough to hit it on the nose, then $x_1 = A/x_1 = \sqrt{A}$. But such good fortune is unlikely.

Otherwise, there are two possibilities. If x_1 is too large, that is, $x_1 > \sqrt{A}$, then A/x_1 is too small, because $x_1 > \sqrt{A}$ implies that $1/x_1 < 1/\sqrt{A}$, and multiplication of this last inequality by A yields

$$\frac{A}{x_1} < \frac{A}{\sqrt{A}} = \sqrt{A}.$$

Similarly, if x_1 is too small, $x_1 < \sqrt{A}$, it follows that $A/x_1 > \sqrt{A}$, so A/x_1 is too large.

Thus in either case one of the two numbers x_1 and A/x_1 is smaller than the true value A and the other is larger. The Babylonians' idea was that we should get a better approximation to $\sqrt{A}$ by *averaging* x_1 and A/x_1, so let us write

$$x_2 = \frac{1}{2}\left(x_1 + \frac{A}{x_1}\right). \tag{1}$$

But now why not repeat the process—treat x_2 as we did our first estimate x_1, and calculate a still better approximation x_3 by averaging again,

$$x_3 = \frac{1}{2}\left(x_2 + \frac{A}{x_2}\right). \tag{2}$$

Indeed, we can repeat this averaging process again and again to generate a sequence $x_1, x_2, x_3, x_4, \ldots$ of presumably better and better approximations to $\sqrt{A}$. Having calculated the nth approximation x_n, we calculate the next one by means of the formula

$$x_{n+1} = \frac{1}{2}\left(x_n + \frac{A}{x_n}\right). \tag{3}$$

which averages x_n and A/x_n.

Equation (3) is an example of an *iterative formula*. Once we have used it to calculate one approximation, we plow this one back into the right-hand side of (3) to calculate the next approximation. It is interesting to note that the words "iteration" and "iterative" stem from the Latin verb *iterare*, to plow again. Much of the usefulness of the computer in computational mathematics stems from its effectiveness (when properly programmed) in carrying out iterative processes like the Babylonian square root method.

To illustrate Formula (3) with $A = 2$, and thus to approximate $\sqrt{2}$, we start with the first guess $x_1 = 1$. Then successive application of (3) yields

$$x_2 = \frac{1}{2}\left(1 + \frac{2}{1}\right) = \frac{3}{2} = 1.5,$$

$$x_3 = \frac{1}{2}\left(\frac{3}{2} + \frac{2}{3/2}\right) = \frac{17}{12} \approx 1.416666667,$$

$$x_4 = \frac{1}{2}\left(\frac{17}{12} + \frac{2}{17/12}\right) = \frac{577}{408} \approx 1.414215686,$$

$$x_5 = \frac{1}{2}\left(\frac{557}{408} + \frac{2}{577/408}\right) = \frac{665,857}{470,832} \approx 1.414213562,$$

rounding off the results to nine decimal places. It happens that x_5 gives $\sqrt{2}$

accurate to all nine decimal places! This is pretty good evidence that the Babylonians were onto something.

To try to see how and why, suppose that in applying Formula (3) iteratively we reach the point that $x_{n+1} = x_n$ to the number of decimal places we are carrying. (Indeed, it can be proved that this always happens if we persist long enough, whatever A and our first estimate x_1 are, as long as both are positive.) Then, replacing x_{n+1} with x_n in (3) and writing $\approx$ to denote equality to an appropriate number of decimal places, we find that

$$x_n \approx \frac{1}{2}\left(x_n + \frac{A}{x_n}\right),$$

$$2x_n \approx x_n + \frac{A}{x_n},$$

$$x_n \approx \frac{A}{x_n},$$

$$x_n^2 \approx A.$$

Thus $x_n \approx \sqrt{A}$ to high accuracy when x_n and x_{n+1} agree to sufficiently many decimal places.

In particular, we can use the Babylonian iteration to compute square roots with greater accuracy than the seven significant figures provided by the SQR function in the version of BASIC that comes with the operating system PC-DOS 1.00 or 1.10. Version 2 of IBM BASIC (with DOS 2.00) supports square roots and the BASIC transcendental functions in double precision (16 significant figures), but with version 1 only the four arithmetical operations can be carried out in double precision. Program SQROOT shown in Listing 1.19 is a double-precision implementation of the Babylonian iteration for computing square roots—line 160 declares all variables beginning with A or X to be double precision (16 significant figures).

The number A whose square root we wish to compute is input at line 170. On the nth pass through the loop in lines 220–260, XOLD denotes x_n and XNEW denotes x_{n+1}. Note in line 210 that $x_1 = A$. Thus our first guess at A is the number A itself.

Line 230 is the Babylonian averaging formula. The values of x_{n+1} and x_n are compared in line 240, and the iteration terminates when the *relative* difference $|x_{n+1} - x_n|/x_n$ is less then 10^{-16}. We use relative rather than absolute differences to prevent loss of significant figures when the numbers involved are very small. Figure 1.20 shows the results of two runs of Program SQROOT, first with $A = 400$ and then with $A = 2$.

```
100 REM--Program SQROOT
110 REM--Applies the Babylonian iteration method
115 REM--to approximate the square root of the
120 REM--input positive number A.
130 REM
140 REM--Declare double precision and input number:
150 REM
160      DEFDBL A, X
170      INPUT "POSITIVE NUMBER"; A
180 REM
190 REM--Babylonian iteration:
200 REM
210      XOLD = A : N = 1
220      PRINT N, XOLD
230      XNEW = .5*(XOLD + A/XOLD)
240      IF ABS(XNEW - XOLD) < XOLD*1E-16
              THEN GOTO 280
250      XOLD = XNEW : N = N + 1
260      GOTO 220
270 REM
280      PRINT "SQUARE ROOT = "; XNEW
290      END
```

Listing 1.19   Program SQROOT

```
RUN
POSITIVE NUMBER? 400
    1              400
    2              200.5
    3              101.247506234414
    4              52.59911041180492
    5              30.10190088122235
    6              21.69504912358706
    7              20.06621767747577
    8              20.00010925778043
    9              20.00000000029843
   10              20
SQUARE ROOT =     20


RUN
POSITIVE NUMBER? 2
    1              2
    2              1.5
    3              1.416666666666667
    4              1.41421568627451
    5              1.41421356237469
    6              1.414213562373095
SQUARE ROOT =     1.414213562373095
```

Figure 1.20   Two runs of Program SQROOT

The Method of Successive Substitutions

The Babylonian square root method is a special case of a successive substitutions method for solving equations of the form

$$x = g(x). \tag{4}$$

The manipulations

$$x^2 = A,$$

$$x = \frac{A}{x},$$

$$2x = x + \frac{A}{x},$$

$$x = \frac{1}{2}\left(x + \frac{A}{x}\right) \tag{5}$$

show that the equation $x^2 = A$ can be written in the form (4) with $g(x) = (1/2)(x + A/x)$. Comparing (3) and (5), we see that in this case the Babylonian averaging Formula (3) takes the form

$$x_{n+1} = g(x_n). \tag{6}$$

The *method of successive substitutions* for solving equation (4) consists of starting with an initial guess x_1, and then calculating $x_2, x_3, \ldots$ in turn by means of formula (6). If we reach the point that $x_{n+1} = x_n$ to the number of significant figures we are using, then $x_n = g(x_n)$, so x_n is a solution of Equation (4). However, with a particular equation $x = g(x)$ and a particular first guess x_1, the iteration may or may not "converge." The pragmatic approach to the method of successive substitutions is simply not to worry about convergence—go ahead and try it. If it works, fine. If not, another method must be used.

Any given equation $f(x) = 0$ that we wish to solve can be rewritten in the form $x = g(x)$ in many different ways. How we want to do it—in preparation for the method of successive substitutions—is our choice. For instance, suppose that we want to solve (surprise!) by successive substitutions the cork ball equation,

$$f(x) = x^3 - 3x^2 + 1 = 0, \tag{7}$$

which we already know has the approximate solutions $-0.5321, 0.6527$, and 2.8794. We can rewrite Equation (7) in the form $x = g(x)$ in the following ways:

$$x = \pm\sqrt{\frac{x^3 + 1}{3}} \tag{8}$$

by "solving" for x^2 and taking the square root;

$$x = (3x^2 - 1)^{1/3} \tag{9}$$

by "solving" for x^3 and taking the cube root; and

$$x = \frac{1}{3x - x^2} \tag{10}$$

by writing $x(x^2 - 3x) = -1$ and "solving" for x.

```
100 REM--Program SUCCSUB
110 REM--Applies the method of successive substitutions
120 REM--to attempt to solve the equation  x = g(x).
130 REM--The function g(x) must be defined in line 170.
140 REM
150 REM--Initialization:
160 REM
170     DEF FNG(X) = SQR((X^3 + 1)/3)
180     INPUT "INITIAL GUESS X1"; XOLD
190     N = 1
200     PRINT "N", "  X" : PRINT
210 REM
220 REM--Iteration:
230 REM
240     PRINT N, XOLD
250     XNEW = FNG(XOLD)
260     IF ABS(XNEW - XOLD) < .00001*ABS(XOLD)
            THEN GOTO 310
270     XOLD = XNEW : N = N + 1
280     IF N > 15 THEN PRINT
            "NO SOLUTION FOUND" : STOP
290     GOTO 240
300 REM
310     PRINT "SOLUTION = "; INT(XNEW*10000)/10000
320     END
```

Listing 1.21   Program SUCCSUB

Listing 1.21 shows Program SUCCSUB, which implements the method of successive substitutions for Equation (8) with the positive sign. It can be applied to any equation of the form $x = g(x)$ after editing $g(x)$ into line 170. Line 180 requests input of our initial guess x_1. The iteration in lines 240–290 terminates if and when the numerical difference between XNEW $= x_{n+1}$ and XOLD $= x_n$ is less than $0.00001 |x_n|$. It often happens that the iterates generated by successive substitution either diverge—do not converge—or converge very slowly. Therefore, we have inserted an additional "stopping condition" in line 280, so that the process will stop after 15 iterations, whether or not a solution has been found. Line 310 directs that only four significant figures of the solution (if any) be displayed.

Figure 1.22 shows the results of three runs of Program SUCCSUB. With the initial guesses $x_1 = 1$ and $x_1 = 2$ the solution $x = 0.6527$ is found, but with $x_1 = 3$ divergence occurs and no solution is found. This suggests that the method will not yield the solution 2.8794 or the solution -0.5321, unless perhaps we take the negative sign in (8). In the problems below we ask you to investigate what happens with other initial guesses, as well as with alternative forms of the cork ball equation.

```
RUN
INITIAL GUESS X1? 1
  N                   X

  1                  1
  2                   .8164966
  3                   .7174796
  4                   .6756089
  5                   .6603988
  6                   .6552398
  7                   .6535341
  8                   .652975
  9                   .6527922
 10                   .6527326
 11                   .6527131
SOLUTION =   .6527


RUN
INITIAL GUESS X1? 2
  N                   X

  1                  2
  2                  1.732051
  3                  1.437145
  4                  1.15011
  5                   .9167532
  6                   .7682171
  7                   .696029
  8                   .6676315
  9                   .6576687
 10                   .6543343
 11                   .653237
 12                   .6528778
 13                   .6527605
 14                   .6527223
 15                   .6527098
SOLUTION =   .6527


RUN
INITIAL GUESS X1? 3
  N                   X

  1                  3
  2                  3.05505
  3                  3.136548
  4                  3.258689
  5                  3.445005
  6                  3.736555
  7                  4.209877
  8                  5.020355
  9                  6.520043
 10                  9.629351
 11                 17.26148
 12                 41.40936
 13                153.8474
 14               1101.729
 15              21113.06
NO SOLUTION FOUND
```

Figure 1.22   Three runs of Program SUCCSUB for Equation (8)

Finally, let us apply the method of successive substitutions to investigate the meaning of the (infinitely) *continued fraction*

$$x = \cfrac{1}{1 + \cfrac{1}{1 + \cfrac{1}{1 + \cfrac{1}{1 + \cdots}}}}. \tag{11}$$

If this symbol has any meaning at all, the number x should satisfy the equation

$$x = \frac{1}{1 + x}. \tag{12}$$

```
RUN
INITIAL GUESS X1? 1
   N                    X

   1                    1
   2                    .5
   3                    .6666667
   4                    .6
   5                    .6250001
   6                    .6153846
   7                    .6190476
   8                    .6176471
   9                    .6181818
  10                    .6179775
  11                    .6180556
  12                    .6180258
  13                    .6180371
SOLUTION =   .61803
```

Figure 1.23   Run of Program SUCCSUB with $g(x) = 1/(1 + x)$ to compute continued fraction

Figure 1.23 shows the results of a run of Program SUCCSUB with $g(x) = 1/(1 + x)$, as in Equation (12). We have altered line 310 so as to display five significant figures of the solution, which is $x = 0.61803$. To explain this result, we can rewrite (12) as $x^2 + x - 1 = 0$, and use the quadratic formula to find the positive solution

$$x = \frac{-1 + \sqrt{5}}{2} \approx 0.61803.$$

PROBLEMS

1. Check the performance of Program SQROOT in computing square roots of very small and very large numbers by running it with inputs ranging from 10^{-10} to 10^{10}.

2. Write a program that uses SQROOT as a subroutine to print a table of square roots of the positive integers from $n = 1$ to 100 (or whatever). *Suggestion:* Load Program SQROOT and execute the command RENUM 500 to renumber its lines starting at 500. Also replace the END in its last line with RETURN. Then insert a "front-end" main program in the form of a FOR-NEXT loop that prints A and $\sqrt{A}$ for $A = 1$ to 100, computing $\sqrt{A}$ by a GOSUB 500.

3. Show that the equation $x^3 = A$ can be written in the form

$$x = \frac{1}{3}\left(2x + \frac{A}{x^2}\right).$$

Then run Program SUCCSUB with $g(x) = (1/3)(2x + A/x^2)$ to compute some cube roots. This amounts to using the iterative formula

$$x_{n+1} = \frac{1}{3}\left(2x_n + \frac{A}{x_n^2}\right),$$

which calculates x_{n+1} as a weighted average of x_n and A/x_n^2.

4. Run Program SUCCSUB for Equation (8) with the positive sign, with a sufficient variety of initial guesses x_1 to convince yourself that the iterates converge to $x = 0.6527$ if x_1 is less than the larger solution 2.8794, but diverges (no solution) if $x_1 > 2.8794$.

5. Investigate the performance of Program SUCCSUB— with a variety of initial guesses—for Equation (8) with the *negative* sign.

6. Run Program SUCCSUB (with various initial guesses) for the alternative form (9) of the cork ball equation. This form yields the largest root 2.8794, but the convergence is very slow, so you should replace 15 in the stopping condition (line 280) by 50 or 100. Because BASIC does not accept fractional powers of negative numbers, you should write $g(x)$ as

$$\text{SGN}(3 * X * X - 1) * (\text{ABS}(3 * X * X - 1))^{\wedge}(1/3)$$

7. Investigate the performance of Program SUCCSUB—with various initial guesses—with the alternative form (10) of the cork ball equation.

8. This problem deals with the cubic equation $x^3 - 3x - 1 = 0$.
   (a) Write the equation in the form $x = 1/(x^2 - 3)$ and use SUCCSUB to find one root, r_1.
   (b) Write the equation in the form $x = (3 + 1/x)^{1/2}$ and use SUCCSUB to find a second root, r_2.
   (c) Divide the polynomial $(x - r_1)(x - r_2)$ into the polynomial $x^3 - 3x - 1$ to find the third root, r_3.

9. Use Program SUCCSUB with $g(x) = 1/(2 + x)$ to find the value of the continued fraction

$$x = \cfrac{1}{2 + \cfrac{1}{2 + \cfrac{1}{2 + \cfrac{1}{2 + \cdots}}}}.$$

Then show that the exact value is $x = \sqrt{2} - 1$.

10. Use Program SUCCSUB with $g(x) = \sqrt{3 + x}$ to investigate the value of the "continued square root"

$$x = \sqrt{3 + \sqrt{3 + \sqrt{3 + \sqrt{3 + \cdots}}}}.$$

Then deduce from the equation $x = \sqrt{3 + x}$ that the exact value is $x = (1 + \sqrt{13})/2$.

11. Vieta's formula is the "infinite product"

$$\frac{2}{\pi} = \sqrt{\frac{1}{2}} \sqrt{\frac{1}{2} + \frac{1}{2}\sqrt{\frac{1}{2}}} \sqrt{\frac{1}{2} + \frac{1}{2}\sqrt{\frac{1}{2} + \frac{1}{2}\sqrt{\frac{1}{2}}}} \cdots,$$

that is,

$$\frac{2}{\pi} = a_1 a_2 a_3 \cdots a_n \cdots,$$

where

$$a_1 = \sqrt{\frac{1}{2}} \qquad \text{and} \qquad a_{n+1} = \sqrt{\frac{1}{2}(1 + a_n)}$$

for $n \geq 1$. Write a program to compute $a_1 a_2 \cdots a_n$ for $n = 1, 2, 3, \ldots$, stopping when $a_n = 1$ to 10 decimal places. What 10-place value for π does your computation yield?

12. Compute

$$x = \sqrt[3]{4 + \sqrt[3]{4 + \sqrt[3]{4 + \cdots}}}$$

accurate to five decimal places. Of what cubic equation is x a positive root?

1.5 NEWTON'S METHOD FOR POLYNOMIALS

In the late 1660s Isaac Newton discovered a remarkably powerful method for approximating solutions of equations. In the case of an equation of the form $x^2 - A = 0$, this method reduces to the Babylonian square root iteration of Section 1.4.

To describe Newton's method for more general equations, let us start with a cubic equation of the form

$$f(x) = ax^3 + bx^2 + cx + d = 0. \tag{1}$$

Let $x = r$ be an actual (but as yet unknown) solution of (1) that we wish to approximate. If x_1 is our initial guess, we write

$$r = x_1 + E, \tag{2}$$

so E denotes the (unknown) error in our guess. Then

$$f(r) = f(x_1 + E) = 0,$$

so substitution of $r = x_1 + E$ in (1) yields

$$a(x_1^3 + 3x_1^2 E + 3x_1 E^2 + E^3)$$
$$+ b(x_1^2 + 2x_1 E + E^2) + c(x_1 + E) + d = 0.$$

Collecting powers of E, we get

$$(ax_1^3 + bx_1^2 + cx_1 + d)$$
$$+ E(3ax_1^2 + 2bx_1 + c) + E^2(3ax_1 + aE + b) = 0. \tag{3}$$

Now suppose that our initial error E is sufficiently small that its square E^2 is negligible. If we ignore the E^2 terms in (3), we get

$$(ax_1^3 + bx_1^2 + cx_1 + d) + E(3ax_1^2 + 2bx_1 + c) \approx 0,$$

so

$$E \approx -\frac{ax_1^3 + bx_1^2 + cx_1 + d}{3ax_1^2 + 2bx_1 + c}, \tag{4}$$

assuming that the denominator is nonzero.

Note that the numerator in (4) is simply the value at x_1 of the original function $f(x)$ in (1). Those who have studied calculus will recognize the denominator in (4) as the value at x_1 of the *derivative*

$$f'(x) = 3ax^2 + 2bx + c \tag{5}$$

of the cubic polynomial $f(x)$. For our present purpose we can take (5) as the *definition* of the new function $f'(x)$, which is called the derivative of $f(x)$. Then (4) can be rewritten as

$$E \approx -\frac{f(x_1)}{f'(x_1)}, \tag{6}$$

and substitution of this value of E in (2) yields

$$r \approx x_1 - \frac{f(x_1)}{f'(x_1)}. \tag{7}$$

The right-hand side in (7) is a correction of our initial guess x_1, and we take it as our second approximation,

$$x_2 = x_1 - \frac{f(x_1)}{f'(x_1)}. \tag{8}$$

This is the beginning of an iterative computation. Once the nth approximation x_n has been calculated, the $(n + 1)$st approximation to our solution r of Equation (1) is

$$x_{n+1} = x_n - \frac{f(x_n)}{f'(x_n)}. \tag{9}$$

Newton's method consists of using this iterative formula to calculate the successive approximations $x_1, x_2, x_3, \dots$, starting with an initial guess x_1.

Example 1

In the case of the familiar cork ball equation,

$$f(x) = x^3 - 3x^2 + 1 = 0, \tag{10}$$

we have $a = 1, b = -3, c = 0$, and $d = 1$, so equation (5) yields the derivative

$$f'(x) = 3(1)x^2 + 2(-3)x + (0) = 3x^2 - 6x. \tag{11}$$

Hence the iterative formula (9) takes the form

$$x_{n+1} = x_n - \frac{x_n^3 - 3x_n^2 + 1}{3x_n^2 - 6x_n}. \tag{12}$$

Listing 1.24 shows Program NEWTON, which implements this iteration. During the nth iteration, X denotes x_n and XNEW denotes x_{n+1}. The values FCTN $= f(x_n)$ and DERIV $= f'(x_n)$ are calculated in lines 230 and 240. If DERIV $= 0$, then we would encounter division by zero in line 260 (the iterative formula), so line 250 prompts us to restart the iteration with a new initial guess x_1.

Figure 1.25 shows the results of three successive runs of Program NEWTON with the initial guesses $x_1 = 1$, $x_1 = -1$, and $x_1 = 3$. Although these are not particularly good initial guesses, we see rapid convergence to the three familiar solutions of the cork ball equation.

```
100 REM--Program NEWTON
105 REM--Applies Newton's method to solve the
110 REM--equation f(x) = 0.  Lines 230 and 240
120 REM--must be edited to read FCTN = f(x) and
130 REM--DERIV = f'(x) respectively. The initial
140 REM--guess is input in line 160.
150 REM
160      INPUT "INITIAL GUESS X1"; X
170      N = 1
180      PRINT "N", "  X" : PRINT
190 REM
200 REM--Newton's iteration:
210 REM
220      PRINT N, X
230      FCTN = X^3 - 3*X*X + 1
240      DERIV = 3*X*X - 6*X
250      IF DERIV = 0 THEN PRINT
         "ZERO DERIVATIVE--TRY NEW INITIAL GUESS" :
         GOTO 160
260      XNEW = X - FCTN/DERIV
270      IF ABS(XNEW - X) < .000001*ABS(X) THEN
         PRINT "SOLUTION = ";XNEW : STOP
280      X = XNEW : N = N + 1
290      GOTO 220
300 REM
310      END
```

Listing 1.24 Program NEWTON

```
RUN
INITIAL GUESS X1? 1
 N                X

 1                1
 2                .6666666
 3                .6527778
 4                .6527036
SOLUTION =  .6527036


RUN
INITIAL GUESS X1? -1
 N                X

 1                -1
 2                -.6666666
 3                -.5486111
 4                -.5323902
 5                -.532089
SOLUTION = -.532089


RUN
INITIAL GUESS X1? 3
 N                X

 1                3
 2                2.888889
 3                2.879452
 4                2.879385
SOLUTION =  2.879385
```

Figure 1.25 Three runs of Program NEWTON to solve the corkball equation

As a general rule, Newton's method converges more rapidly than any of the iterative methods discussed previously. Indeed, it is frequently asserted that each iteration of Newton's method doubles the number of decimal places of accuracy. Although this is not always true, it happens often enough for the statement to be a reasonable rule of thumb. On the other hand, there are equations such that (for certain initial guesses) the iterates produced by Newton's method diverge spectacularly. In the exceptional case where the solution r of $f(x) = 0$ is also a solution of the equation $f'(x) = 0$, the convergence is quite slow—this happens when r is a multiple root of $f(x)$.

To apply Program NEWTON to solve a new equation $f(x) = 0$, we need only edit the function $f(x)$ and its derivative $f'(x)$ into lines 230 and 240. Of course, we must know how to calculate $f'(x)$. If $f(x)$ is an nth-degree polynomial, our equation is of the form

$$f(x) = a_n x^n + a_{n-1}x^{n-1} + \cdots + a_1 x + a_0 = 0. \qquad (13)$$

Then a generalization of our analysis of the cubic equation (1) leads to Newton's iterative formula

$$x_{n+1} = x_n - \frac{f(x_n)}{f'(x_n)} \qquad (9)$$

with the derived function

$$f'(x) = na_n x^{n-1} + (n-1)a_{n-1}x^{n-2} + \cdots + 2a_2 x + a_1. \qquad (14)$$

Calculus students will observe that $f'(x)$ is the "ordinary" derivative of the polynomial $f(x)$. That is, $f'(x)$ is obtained from $f(x)$ if we multiply each term by the exponent of x in that term, and then decrease the exponent by one (the constant term a_0 simply disappears). For instance, if

$$f(x) = 2x^4 - 3x^3 + 7x + 10,$$

then

$$f'(x) = (4)(2x^{4-1}) + (3)(-3x^{3-1}) + (1)(7x^{1-1})$$
$$= 8x^3 - 9x^2 + 7.$$

Example 2

For the equation $f(x) = x^2 - A = 0$, the derivative is $f'(x) = 2x$, so Newton's iterative formula is

$$x_{n+1} = x_n - \frac{x_n^2 - A}{2x_n}$$

$$= \frac{x_n^2 + A}{2x_n}$$

$$x_{n+1} = \frac{1}{2}\left(x_n + \frac{A}{x_n}\right).$$

Thus we obtain the Babylonian square root iteration as a special case of Newton's method.

Example 3

For the equation $f(x) = x^3 - A = 0$, the solution of which is the cube root $\sqrt[3]{A}$, the derivative is $f'(x) = 3x^2$, so (9) yields

$$x_{n+1} = x_n - \frac{x_n^3 - A}{3x_n^2}$$

$$= \frac{1}{3}\left(2x_n + \frac{A}{x_n^2}\right). \tag{15}$$

Hence if we replace line 230 in Program SQROOT (Listing 1.19) with

$$230 \quad XNEW = (2*XOLD + A/XOLD^2)/3,$$

we get a program that computes *cube* roots.

To write a program to solve a general polynomial equation upon input of the coefficients, it is convenient to have a more efficient way of evaluating the polynomial

$$f(x) = a_n x^n + a_{n-1} x^{n-1} + \cdots + a_1 x + a_0 \tag{16}$$

at $x = c$ than merely substituting and calculating directly the powers of c that appear in $f(c)$. For this purpose we rewrite (16) as

$$f(c) = a_0 + a_1 c + a_2 c^2 + \cdots + a_{n-1} c^{n-1} + a_n c^n$$
$$= a_0 + c(a_1 + c(a_2 + c(\cdots + c(a_{n-1} + ca_n))\cdots)).$$

This is the *stacked form* of the polynomial. We can unravel it from the inside out by writing

$$b_n = a_n,$$
$$b_{n-1} = a_{n-1} + cb_n,$$
$$b_{n-2} = a_{n-2} + cb_{n-1},$$
$$\vdots \tag{17}$$
$$b_1 = a_1 + cb_2,$$
$$b_0 = a_0 + cb_1 = f(c).$$

Thus the values $b_{n-1}, b_{n-2}, \ldots, b_1, b_0$ are calculated in descending order of subscripts by means of the recurrence relation

$$b_k = a_k + cb_{k+1} \quad (k \leq n - 1), \tag{18}$$

starting with $b_n = a_n$. The final value b_0 is the value of $f(c)$. This is *Horner's method* of calculating $f(c)$.

Furthermore, it is true (see Problem 13) that

$$\frac{f(x)}{x - c} = b_n x^{n-1} + b_{n-1} x^{n-2} + \cdots + b_2 x + b_1 + \frac{b_0}{x - c}. \qquad (19)$$

That is, if the polynomial $f(x)$ is divided by $(x - c)$, then *the quotient is the polynomial*

$$q(x) = b_n x^{n-1} + b_{n-1} x^{n-2} + \cdots + b_2 x + b_1$$

with remainder b_0. The computation of the b_k values is often presented as the following scheme:

a_n	a_{n-1}	a_{n-2}	$\cdots$	a_1	a_0
	$+cb_n$	$+cb_{n-1}$	$\cdots$	$+cb_2$	$+cb_1$
b_n	b_{n-1}	b_{n-2}	$\cdots$	b_1	b_0.

In this form the algorithm is known as *synthetic division*. For instance, to divide $x - 2$ into $f(x) = 2x^3 - x^2 - 11x + 10$, we carry out the following computation, multiplying each successive b_k by $c = 2$ and adding the next a_k:

a_k:	2	-1	-11	10
		4	6	-10
b_k:	2	3	-5	0.

The final zero means that $f(2) = 0$, so $x - 2$ is a factor of $f(x)$. The preceding values $b_3 = 2, b_2 = 3, b_1 = -5$ tell us that the quotient is $q(x) = 2x^2 + 3x - 5$, so $f(x)$ factors as

$$2x^3 - x^2 - 11x + 10 = (x - 2)(2x^2 + 3x - 5).$$

Consequently, if we were attempting to solve the equation

$$2x^3 - x^2 - 11x + 10 = 0,$$

and somehow found first the solution $x = 2$, we could then, by synthetic division, *deflate* this cubic equation to the quadratic equation

$$2x^2 + 3x - 5 = 0$$

with two solutions $x = 1$ and $x = -5/2$ that are easily found (by factoring).

```
100 REM--Program HORNER
105 REM--When the degree n, the number c, and
110 REM--the coefficients An,---,A1,A0 of a
115 REM--polynomial f(x) are input, program
120 REM--computes by Horner's method the
125 REM--coefficients Bn,---,B2,B1 of its
130 REM--quotient by (x - c), and the remainder
140 REM--B0 = f(c).
150 REM
160 REM--Input degree, divisor, and coefficients:
170 REM
180        INPUT "DEGREE N = "; N
190        DIM A(N), B(N)
200        INPUT "NUMBER C = "; C
210        FOR K = N TO 0 STEP -1
220            PRINT "A("; K; ") = " ;
230            INPUT A(K)
240        NEXT K
250        PRINT "K", "AK", "BK" : PRINT
260 REM
270 REM--Horner's algorithm:
280 REM
290        B(N) = A(N)
300        PRINT N, A(N), B(N)
310        FOR K = N-1 TO 0 STEP -1
320            B(K) = A(K) + C*B(K+1)
330            PRINT K, A(K), B(K)
340        NEXT K
350        PRINT : PRINT "f("; C; ") = "; B(0)
360        END
```

Listing 1.26   Program HORNER

```
RUN
DEGREE N = ?  3
NUMBER C = ?  2
A( 3 ) = ?  2
A( 2 ) = ? -1
A( 1 ) = ? -11
A( 0 ) = ?  10
   K              AK              BK

   3               2               2
   2              -1               3
   1             -11              -5
   0              10               0

f( 2 ) =  0
```

Figure 1.27   Run of Program HORNER

Listing 1.26 shows Program HORNER, which implements this method. The coefficients of the nth-degree polynomial $f(x)$ are input and placed in the one-dimensional array $A(K), K = 0, 1, \ldots, N$, with $a_k = A(K)$ denoting the coefficient of x^k. Line 320 implements the recurrence relation (18), and the coefficients of the quotient polynomial are placed in the array $B(K)$. The coefficients of both $f(x)$ and $q(x)$ as well as the value $f(c) = B(0)$ are then printed. Figure 1.27 shows a run of Program HORNER corresponding to the simple synthetic division exhibited above.

```
100 REM--Program POLYNEWT
110 REM--Solves by Newton's method a general
115 REM--polynomial equation f(x) = 0 whose
120 REM--coefficients A(N), A(N-1),---,A(1),A(0)
125 REM--are input.
130 REM
140 REM--Input degree and coefficients:
150 REM
160       INPUT "DEGREE N= "; N
170       DIM  A(N), AD(N), B(N)
180       FOR K = N TO 0 STEP -1
190           PRINT "A(";K;") = ";
200           INPUT A(K)
210       NEXT K
220 REM
230 REM--Derivative coefficients and initial guess:
240 REM
250       FOR K = N - 1 TO 0 STEP -1
260           AD(K) = (K+1)*A(K+1)
270       NEXT K
280       INPUT "INITIAL GUESS X1"; X
290       PRINT " M", "  XM", " f(XM)" : PRINT
300       M = 1
310 REM
320 REM--Newton's iteration:
330 REM
340       B(N) = A(N)
350       FOR K = N-1 TO 0 STEP -1
360           B(K) = A(K) + X*B(K+1)
370       NEXT K
380       FCTN = B(0)
390       PRINT M, X, FCTN
400       B(N-1) = AD(N-1)
410       FOR K = N-2 TO 0 STEP -1
420           B(K) = AD(K) + X*B(K+1)
430       NEXT K
440       DERIV = B(0)
450       IF DERIV = 0 THEN PRINT
          "ZERO DERIVATIVE--TRY NEW INITIAL GUESS":
          GOTO 280
460       XNEW = X - FCTN/DERIV
470       IF ABS(XNEW - X) < .000001*ABS(X) THEN
          PRINT: PRINT "SOLUTION = " ; XNEW :   STOP
480       X = XNEW : M = M+1
490       GOTO 340
500       END
```

Listing 1.28   Program POLYNEWT

We are now prepared to apply Newton's method to the general polynomial equation (13) with coefficients $a_n, \ldots, a_1, a_0$ (in decreasing order of degree). Listing 1.28 shows Program POLYNEWT, which was written for this purpose. After the degree n is input in line 160, the loop in lines 180–210 inputs the coefficients and places them in an array A(K), K = 0, 1, 2, ..., N. The loop in lines 250–270 then calculates the coefficients of the derivative (14) and places them in an array AD(K); note that the coefficient of x^k in $f'(x)$ is $(k + 1)a_{k+1}$. The loop in lines 340–370 applies Horner's method to evaluate FCTN $= f(x)$, while the loop in lines 400–430 similarly evaluates DERIV $= f'(x)$. Newton's iteration itself appears in lines 450–490, which are the same as lines 250-290 of Program NEWTON. In case the iterates $x_1, x_2, \ldots, x_m, \ldots$ do not converge, you should be ready to hit the Control-Break keys and try a new initial guess (GOTO 280).

Complete Solution of a Cubic Equation

Finally, suppose that we want to find (at least approximately) all three solutions of the cubic equation

$$f(x) = a_3 x^3 + a_2 x^2 + a_1 x + a_0 = 0. \tag{20}$$

We may assume that $a_3 > 0$. Then $f(x) < 0$ if x is a sufficiently large negative number, while $f(x) > 0$ if x is a sufficiently large positive number. It follows that the cubic equation always has at least one real solution $x = r$. Once we have found it, we can use Program HORNER to deflate the cubic equation to a quadratic equation, and then solve the latter to find the other two solutions. (For simplicity in the discussion that follows, we will ignore the roundoff error that deflation of a polynomial equation entails.)

```
100 REM--Program QUADRAT
110 REM--Program computes roots of a quadratic equation.
120 REM
130 REM--Input coefficients and compute discriminant:
140 REM
150       INPUT "COEFFICIENTS A,B,C"; A,B,C
160       LET D = B*B - 4*A*C
170       IF D < 0 THEN GOTO 320
180 REM
190 REM--Case of real equal roots:
200 REM
210       IF D = 0 THEN GOTO 220 ELSE GOTO 250
220       PRINT "X1 = X2 = "; -B/(2*A)
230       STOP
240 REM
250 REM--Case of real unequal roots:
260 REM
270       IF D > 0 THEN LET E = SQR(D)
280       PRINT "X1 = "; (-B+E)/(2*A),,
290       PRINT "X2 = "; (-B-E)/(2*A)
300       STOP
310 REM
320 REM--Case of complex roots:
330 REM
340       LET R = -B/(2*A) : I = SQR(-D)/(2*A)
350       PRINT "X1 = "; R; " + "; I; "i"
360       PRINT "X2 = "; R; " - "; I; "i"
370       END
```

Listing 1.29   Program QUADRAT

Program QUADRAT, shown in Listing 1.29, applies the quadratic formula to compute the two roots x_1 and x_2 of a quadratic equation $ax^2 + bx + c = 0$ when the coefficients a, b, c are input. The discriminant $D = b^2 - 4ac$ is calculated in line 160, and the three possibilities—real equal roots, real unequal roots, and conjugate complex roots—correspond to the cases $D = 0$, $D > 0$, and $D < 0$, respectively. Problem 11 below describes a refinement of Program QUADRAT that is sometimes more accurate.

Example 4

Find all three solutions of the cubic equation

$$f(x) = 4x^3 - 42x^2 - 19x - 28 = 0. \tag{21}$$

Solution.   Because $f(0) = -28 < 0$ while $f(x) > 0$ for x sufficiently large, Equation (21) has at least one positive solution $x = r_1$. We could use Program TABULATE to locate this solution in some interval of reasonable length, but instead, let's just run Program POLYNEWT with initial guess $x_0 = 0$.

```
RUN
DEGREE N= ?  3
A( 3 ) = ?   4
A( 2 ) = ?  -42
A( 1 ) = ?  -19
A( 0 ) = ?  -28
INITIAL GUESS X1?  0
 M                 XM                  f(XM)

 1            0                   -28
 2           -1.473684            -104.0152
 3           -.6787677            -35.70479
 4            .1411803            -31.50831
 5           -.8878316            -47.03681
 6           -.1645979            -26.02836
 7           -5.532749            -1886.012
 8           -3.213179            -533.2774
 9           -1.790351            -151.5632
10           -.8980354            -47.70592
11           -.1764495            -25.97708
12           -7.004213            -3329.876
13           -4.128825            -947.0742
14           -2.349907            -267.1837
15           -1.257832            -78.5114
16           -.5146595            -29.89148
17            .5758768            -52.10637
18           -.2460675            -25.92738
19           10.57386             -195.8736
20           11.02469             17.61952
21           10.99037             .1060963
22           10.99016             1.430512E-04

SOLUTION =   10.99016
```

Figure 1.30   Run of Program POLYNEWT for Example 4

The result is shown in Figure 1.30. Note that after 19 iterations it is not apparent that we are getting anywhere. But then the next four iterations yield the solution $r_1 \approx 10.99016$. Since $f(x_7) \approx -1886$ and $f(x_{12}) \approx -3330$, things certainly got worse before they got better! Equation (21) is a rather obstreperous cubic.

```
RUN
DEGREE N = ? 3
NUMBER C = ? 10.99016
A( 3 ) = ?  4
A( 2 ) = ? -42
A( 1 ) = ? -19
A( 0 ) = ? -28
   K              AK               BK

   3              4                4
   2             -42               1.96064
   1             -19               2.547747
   0             -28               1.430512E-04

f( 10.99016 ) =  1.430512E-04
```

Figure 1.31   Deflating the equation of Example 4

Figure 1.31 shows the result of a run of Program HORNER to divide the cubic in (21) by $(x - 10.99016)$. Observe that this deflates (21) to the quadratic equation

$$4x^2 + 1.960640x + 2.547747 = 0. \tag{22}$$

Finally, we run Program QUADRAT to solve this quadratic equation, and get (Figure 1.32) the conjugate complex solutions $-0.24508 \pm 0.7595213i$. Thus the three (approximate) solutions of Equation (21) are $x = 10.99016$ and $-0.24508 \pm 0.7595213i$.

```
RUN
COEFFICIENTS A,B,C? 4, 1.96064, 2.547747
X1 = -.24508  +  .7595213 i
X2 = -.24508  -  .7595213 i
```

Figure 1.32   Run of Program QUADRAT to solve Equation (22)

PROBLEMS

Apply the method of Example 4 to find all three solutions of each of the cubic equations in Problems 1–5.

1. $x^3 - 3x^2 + 1 = 0$

2. $x^3 - 3x^2 + 4 = 0$

3. $x^3 - 3x^2 + 7 = 0$

4. $3x^3 - 9x + 2 = 0$

5. $x^3 = 2$

Apply the method of Example 4 (but deflating twice) to find all four solutions of each of the quartic equations of Problems 6–9.

6. $3x^4 + 4x^3 - 10 = 0$

7. $3x^4 + 4x^3 + 1 = 0$

8. $x^4 + 6x^3 - 110x^2 + 150x + 225 = 0$

9. $x^4 - 16x^3 + 72x^2 - 96x + 24 = 0$

10. Find all five solutions of the fifth-degree equation

$$x^5 + x^4 - 3x^3 - 3x^2 - 3x + 1 = 0.$$

11. The solutions of the quadratic equation

$$x^2 - 10000x + 1 = 0$$

are $x_1 = 9999.9999$ and $x_2 = 0.0001$, accurate to four decimal places. However, Program QUADRAT gives $x_1 = 10,000$ and $x_2 = 0$. This is acceptable for the larger solution x_1, but $x_2 = 0$ is not even accurate to one significant figure. This is a result of cancellation of decimal places in the numerator of the quadratic formula

$$x_2 = \frac{-b \pm \sqrt{b^2 - 4ac}}{2a}.$$

To remedy this, we note from

$$a(x - x_1)(x - x_2) = ax^2 + bx + c$$

that $ax_1x_2 = c$. Alter Program QUADRAT so that the first solution x_1 that is found is the one that is numerically larger, and then $x_2 = c/ax_1$ is calculated. This version of QUADRAT should give $x_1 = 10,000$ and $x_2 = 0.0001$ as the solutions of the quadratic equation above.

12. Following the steps described below, write a program that inputs the four coefficients of the cubic equation

$$f(x) = a_3x^3 + a_2x^2 + a_1x + a_0 = 0$$

and then displays all three solutions.

  1. First we need to locate one root. Let m denote the maximum of the absolute values $|a_2/a_3|$, $|a_1/a_3|$, and $|a_0/a_3|$. According to a well-known elementary estimate due to Cauchy, every root r of the equation satisfies

$$|r| \leq m + 1.$$

     Hence if $f(0) < 0$, there is a solution in $[0, m + 1]$, while if $f(0) > 0$, there is a solution in $[-m - 1, 0]$. Use Program BISECT as a subroutine to find an interval $[c, d]$ of length < 1 that contains a solution.
  2. Next use Program POLYNEWT as a subroutine to find the first (real) solution r_1.
  3. Then use Program HORNER as a subroutine to deflate the cubic equation to a quadratic equation.
  4. Finally, use Program QUADRAT as a subroutine to find the other two solutions.

13. Verify Equation (19) in the text by first multiplying both sides by $x - c$, and then applying the recurrence relation (18).

1.6 COMPUTER GRAPHICS AND EYEBALL SOLUTIONS

To approximate the solution(s) of the equation $f(x) = 0$, it suffices to construct a reasonably accurate picture of the graph $y = f(x)$. Then we can actually "see" the solutions—the points where the curve $y = f(x)$ crosses the x-axis (where $y = 0$). For instance, suppose that $y = f(x)$ looks like the graph shown in Figure 1.33. Then we can locate visually a solution of the equation $f(x) = 0$ between $x = 1.4$ and $x = 1.5$. Indeed, it is obviously closer to 1.4 than to 1.5, so we can see that the indicated solution is $x \approx 1.4$ accurate to one decimal place.

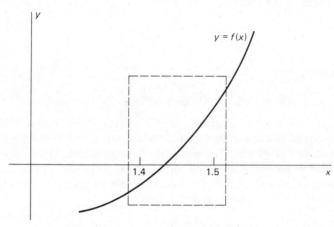

Figure 1.33   Visual location of a root of $f(x) = 0$

Now suppose we could magnify greatly the region within the dashed rectangle in Figure 1.33, and thus "zoom in" on the graph $y = f(x)$ on the interval $[1.4, 1.5]$. Then we could eyeball the solution accurate to two decimal places. Another "zoom" might give three-place accuracy, and so on. In this section we employ the graphics capabilities of the personal computer to solve equations by carrying out successive magnifications of appropriate portions of graphs.

First we must describe the way graphical figures are plotted on the IBM Personal Computer monitor screen. In "high-resolution mode" (executed by the command SCREEN 2) the screen is subdivided into a rectangular pattern of $200 \times 640 = 12{,}800$ individual pixels (or dots). There are 200 horizontal rows and 640 vertical columns of pixels, each of which can be individually turned on (visible) or off (not visible). The rows are numbered 0 to 199 from top to bottom, and the columns are numbered 0 to 639 from left to right, as indicated in Figure 1.34.

In essence, the screen is provided with a uv-coordinate system, with the origin $(0, 0)$ being the upper left-hand corner. The u-coordinate is

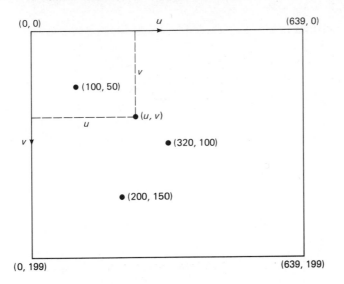

Figure 1.34 The high-resolution monitor screen

measured to the right, from 0 to 639, while the v-coordinate is measured downward, from 0 to 199. The pixel (u, v), where u and v are integers with $0 \leq u \leq 639$ and $0 \leq v \leq 199$, is turned "on" by executing the command

$$\text{PSET} \qquad (\text{u, v}).$$

We can also draw a straight-line segment on the screen from the point (u_1, v_1) to the point (u_2, v_2) by executing the command

$$\text{LINE} \qquad (\text{u}_1, \text{v}_1) - (\text{u}_2, \text{v}_2).$$

These two graphics commands will suffice to construct our graphs.

However, we must contend with the nonstandard orientation of the uv-coordinate system. We are accustomed to an xy-coordinate system in which the x-axis points to the right (like the u-axis), but the y-axis points upward (unlike the v-axis). Suppose that we want to draw on the monitor screen a geometric figure that is located in an ordinary xy-coordinate system (on our sheet of paper, for instance) within the rectangle with lower left vertex (x_{min}, y_{min}) and upper right vertex (x_{max}, y_{max}). This rectangle, shown in Figure 1.35, is called our *window* in the xy-plane.

In particular, suppose that we want to copy this window into the rectangle in the screen's uv-system that has upper left vertex (u_{min}, v_{min}) and lower right vertex (u_{max}, v_{max}). This rectangle, shown in Figure 1.36, is called our *viewport* on the monitor screen. The question is: Given a point (x, y) on the graph in the window, which pixel (u, v) in the viewport should be turned on to represent (x, y)?

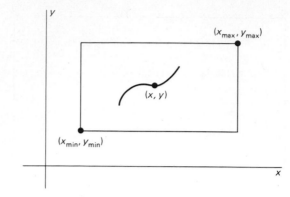

Figure 1.35   A window in the xy-plane

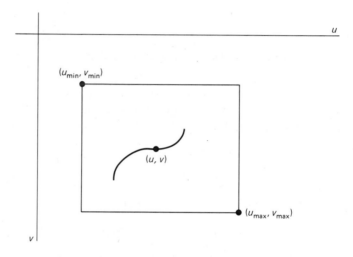

Figure 1.36   A viewport in the screen's uv-plane

In order that objects in the window be portrayed with similar shapes in the viewport, we want to define a linear transformation from xy-coordinates to uv-coordinates. In order that horizontal lines map to horizontal lines, and vertical lines to vertical lines, u should depend only on x and v should depend only on y. Thus the transformation from (x, y) to (u, v) should be described by linear equations of the form

$$u = ax + b, \tag{1}$$
$$v = cy + d, \tag{2}$$

where a, b, c, and d are constants to be determined.

The fact that u_{min} should correspond to x_{min} and u_{max} should correspond to x_{max} means that

$$u_{min} = ax_{min} + b \quad \text{and} \quad u_{max} = ax_{max} + b. \tag{3}$$

These two linear equations are readily solved for a and b in terms of the minimum and maximum values of x and u. When this is done, and the resulting values of a and b are substituted in Equation (1), the result is

$$u = u(x) = u_{min} + \frac{u_{max} - u_{min}}{x_{max} - x_{min}}(x - x_{min}). \tag{4}$$

Similarly, we impose on v the conditions that

$$v_{min} = cy_{max} + d \quad \text{and} \quad v_{max} = cy_{min} + d. \tag{5}$$

(Note the interchange of max and min due to the downward orientation of the v-axis.) When we solve Equations (5) and substitute the resulting values of c and d in Equation (2), we get

$$v = v(y) = v_{max} - \frac{v_{max} - v_{min}}{y_{max} - y_{min}}(y - y_{min}). \tag{6}$$

The transformation $(x, y) \longrightarrow (u(x), v(y))$ defined by Equations (4) and (6) is called the *viewing transformation* from the xy-window of Figure 1.35 to the uv-viewport of Figure 1.36. Once we have chosen our window and our viewport, the values $x_{min}, x_{max}, y_{min}, y_{max}$ and $u_{min}, u_{max}, v_{min}, v_{max}$ are known constants. We can then turn on the pixel corresponding to the point (x, y) by executing the command

PSET   (u(x), v(y)).

With this preparation we can write a program to plot the graph $y = f(x)$ of a given function.

Listing 1.37 shows Program GRAPH, which plots the graph $y = f(x)$ on the input interval [XMIN, XMAX]. We have defined in line 140 the function $f(x) = x^3 - 3x^2 + 1$ corresponding to the familiar cork ball equation. Any other desired function can be edited into line 140.

The purpose of lines 190–260 is to set the scale on the y-axis by choosing for YMIN and YMAX the minimum and maximum values of the function $f(x)$ on the interval [XMIN, XMAX]. For convenience, however, we sample only the values of $f(x)$ at the endpoints of 20 equal subintervals. At this point we have chosen our window in the xy-plane. The screen viewport is defined in lines 300–310, and the viewing transformation corresponding to Equations (4) and (6) is defined in lines 320–330.

The largest single part of the program consists of lines 370–520, which draw and label the axes in the viewport. Line 410 draws the horizontal straight line corresponding to the x-axis, so we can eyeball where the curve $y = f(x)$ crosses it. The loop in lines 420–440 places "ticks" subdividing the interval [XMIN, XMAX] into 10 equal subintervals so that we can readily

```
100 REM--Program GRAPH
110 REM--Plots graph of the function y = f(x)
115 REM--on the input interval [XMIN,XMAX].  The
120 REM--function f(x) must be defined in line 140.
130 REM
140       DEF FNF(X) =  X^3 - 3*X*X + 1
150       INPUT "XMIN, XMAX"; XMIN, XMAX
160 REM
170 REM--Set scale on vertical axis:
180 REM
190       YMAX = FNF(XMIN) : YMIN = FNF(XMIN)
200       FOR I=1 TO 20
210          X = XMIN + I*(XMAX - XMIN)/20
220          IF YMAX < FNF(X) THEN YMAX = FNF(X)
230          IF YMIN > FNF(X) THEN YMIN = FNF(X)
240          IF YMIN > 0 THEN YMIN = 0
250          IF YMAX < 0 THEN YMAX = 0
260       NEXT I
270       DX = XMAX - XMIN : DY = YMAX - YMIN
275 REM
280 REM--Define viewing transformation:
290 REM
300       UMIN = 150 : UMAX = 550  :  DU = 400
310       VMIN = 19  : VMAX = 179  :  DV = 160
320       DEF FNU(X) = UMIN + (X - XMIN)*DU/DX
330       DEF FNV(Y) = VMAX - (Y - YMIN)*DV/DY
340 REM
350 REM--Draw and label axes:
360 REM
370       SCREEN 2 : CLS : KEY OFF
380       LINE (100,   2) - (100,190)
390       LINE ( 95,  19) - (105,  19)
400       LINE ( 95,179) - (105,179)
410       LINE ( 75, FNV(0)) - (625, FNV(0))
420       FOR I=0 TO 10
430          LINE (150 + 40*I, FNV(0) - 5)
                - (150 + 40*I, FNV(0) + 5)
440       NEXT I
450       K = 2 + FNV(0)/8
460       LOCATE  1,14 : PRINT "y"
470       LOCATE  K,78 : PRINT "x"
480       LOCATE  3,5  : PRINT USING "+#.####"; YMAX
490       LOCATE 23,5  : PRINT USING "+#.####"; YMIN
500       LOCATE  K,18 : PRINT XMIN
510       LOCATE  K,43 : PRINT (XMIN + XMAX)/2
520       LOCATE  K,68 : PRINT XMAX
530 REM
540 REM--Plot graph of function:
550 REM
560       X = XMIN : H = (XMAX - XMIN)/200
570       FOR I=0 TO 200
580          Y = FNF(X)
590          PSET (FNU(X), FNV(Y))
600          X = X + H
610       NEXT I
620       LOCATE 1,1
630 REM
640       END
```

Listing 1.37 Program GRAPH

spot the location of the solution. In lines 460–520 we use the LOCATE command to place the cursor for printing of labels. For instance, line 460 prints the character y in the first row and fourteenth column on the screen; the LOCATE command refers to the 25 rows and 80 columns of characters on the screen, rather than to the rows and columns of pixels. One character corresponds to an 8 by 8 block of pixels; this is the origin of the divisor 8 in line 450, which calculates the row in which to print the x-axis labels.

Finally, the graph $y = f(x)$ itself is plotted by the loop in lines 570–610. The point (x, y) with $y = f(x)$ is plotted—for 201 equally spaced values of x— by turning on the pixel at $(u(x), v(y))$ in line 590.

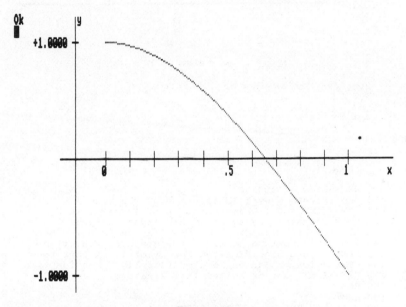

Figure 1.38

Figure 1.38 shows the result of a run of Program GRAPH with XMIN = 0 and XMAX = 1. We see that the graph crosses the x-axis somewhere in the interval $[0.6, 0.7]$, so we run the program again with XMIN = 0.6 and XMAX = 0.7. In Figure 1.39 we see that the graph crosses the x-axis in the interval $[0.65, 0.66]$, but closer to the endpoint 0.65. Thus our solution of the equation $x^3 - 3x^2 + 1 = 0$ is $x \approx 0.65$ accurate to two decimal places.

We could continue to refine our accuracy in this way, but why not write a program that does it automatically, zooming again at the press of a key? Listing 1.40 shows Program ZOOMSOLV, which employs GRAPH as a subroutine. Note that lines 470–920 of ZOOMSOLV are identical to lines 170–620 of GRAPH. To write ZOOMSOLV, we first loaded the latter lines of GRAPH and renumbered them starting at 470. Then we wrote a "front end"

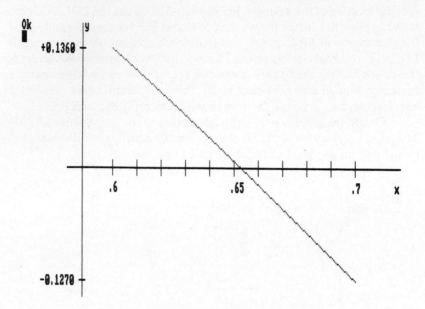

Figure 1.39

```
100 REM--Program ZOOMSOLV
105 REM--Presents eyeball solution of the equation
115 REM--f(x) = 0.  When endpoints of interval [a,b]
120 REM--containing a solution are input, program
130 REM--first plots graph of y = f(x) on [a,b].  Upon
140 REM--each command to ZOOM, the graph is replotted
150 REM--with a magnification factor of 10 so that the
160 REM--solution can be located visually.  The function
170 REM--f(x) must be edited into line 200.
180 REM
190 REM
200        DEF FNF(X) = X^3 - 3*X*X + 1
210        INPUT "ENDPOINTS A,B"; XMIN, XMAX
220        IF FNF(XMIN)*FNF(XMAX) > 0 THEN GOTO 210
230 REM
240        GOSUB 450 :    REM--Plot graph
250 REM
260        INPUT "WANT TO ZOOM? Y or N"; Z$
270        IF Z$ <> "Y" AND Z$ <> "y" THEN 999
280 REM
290 REM--Choose new subinterval:
300 REM
310        X = XMIN : H = (XMAX - XMIN)/10
320        FOR J=1 TO 10
330            IF FNF(X)*FNF(X+H) < 0 THEN GOTO 340
                  ELSE GOTO 350
340            XMIN = X  :  XMAX = X + H  :   GOTO 240
350            X = X + H
360        NEXT J
```

Listing 1.40 Program ZOOMSOLV

```
370 REM
380 REM
450 REM--Subroutine GRAPH:
460 REM
470 REM--Set scale on vertical axis:
480 REM
490        YMAX = FNF(XMIN) : YMIN = FNF(XMIN)
500        FOR I=1 TO 20
510            X = XMIN + I*(XMAX - XMIN)/20
520            IF YMAX < FNF(X) THEN YMAX = FNF(X)
530            IF YMIN > FNF(X) THEN YMIN = FNF(X)
540        IF YMIN > 0 THEN YMIN = 0
550        IF YMAX < 0 THEN YMAX = 0
560        NEXT I
570        DX = XMAX - XMIN : DY = YMAX - YMIN
575 REM
580 REM--Define viewing transformation:
590 REM
600        UMIN = 150 : UMAX = 550  :  DU = 400
610        VMIN = 19  : VMAX = 179  :  DV = 160
620        DEF FNU(X) = UMIN + (X - XMIN)*DU/DX
630        DEF FNV(Y) = VMAX - (Y - YMIN)*DV/DY
640 REM
650 REM--Draw and label axes:
660 REM
670        SCREEN 2 : CLS : KEY OFF
680        LINE (100,  2) - (100,190)
690        LINE ( 95, 19) - (105, 19)
700        LINE ( 95,179) - (105,179)
710        LINE ( 75, FNV(0)) - (625, FNV(0))
720        FOR I=0 TO 10
730            LINE (150 + 40*I, FNV(0) - 5)
                  - (150 + 40*I, FNV(0) + 5)
740        NEXT I
750        K = 2 + FNV(0)/8
760        LOCATE  1,14 : PRINT "y"
770        LOCATE  K,78 : PRINT "x"
780        LOCATE  3,5  : PRINT USING "+#.####"; YMAX
790        LOCATE 23,5  : PRINT USING "+#.####"; YMIN
800        LOCATE  K,18 : PRINT XMIN
810        LOCATE  K,43 : PRINT (XMIN + XMAX)/2
820        LOCATE  K,68 : PRINT XMAX
830 REM
840 REM--Plot graph of function:
850 REM
860        X = XMIN : H = (XMAX - XMIN)/200
870        FOR I=0 TO 200
880            Y = FNF(X)
890            PSET (FNU(X), FNV(Y))
900            X = X + H
910        NEXT I
920        LOCATE 1,1
930 REM
940        RETURN
950 REM
999        END
```

Listing 1.40 (con't.)

main program to call GRAPH as a subroutine. At line 210 we input the endpoints of an interval $[a, b]$ on which $f(x)$ changes sign, so $[a, b]$ contains a solution of the equation $f(x) = 0$. The graph of $y = f(x)$ on $[a, b]$ is then plotted by the subroutine, and the next line 260 asks whether or not we want to zoom. If so, the new subinterval on which $f(x)$ changes sign is selected by the loop in lines 320–360.

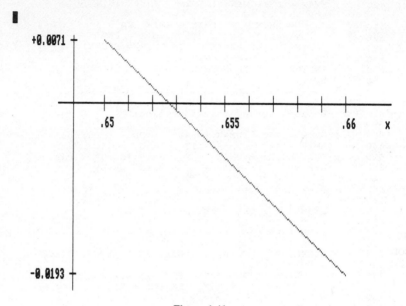

Figure 1.41

When we run Program ZOOMSOLV with $A = 0$, $B = 1$, we first get the graph shown in Figure 1.38. The first zoom gives the graph in Figure 1.39. Two more zooms yield Figures 1.41 and 1.42. From the latter it is evident that the solution of the cork ball equation $x^3 - 3x^2 + 1 = 0$ in the interval $[0, 1]$ is $x \approx 0.6527$ accurate to four decimal places.

Obviously, Program ZOOMSOLV is a powerful graphics tool for solving an equation $f(x) = 0$. We first plot the graph $y = f(x)$ on various initial subintervals (without zooming) to locate the various solutions. Then we zoom in on the graph to eyeball each solution to several decimal places of accuracy.

This brings us to the end of Chapter 1. We have demonstrated the versatility of the personal computer by applying it to the specific problem of solving a given equation, using only quite elementary mathematical techniques. In essence, we are able to employ raw computational power in lieu of more advanced mathematical methods. In the following chapters we will see

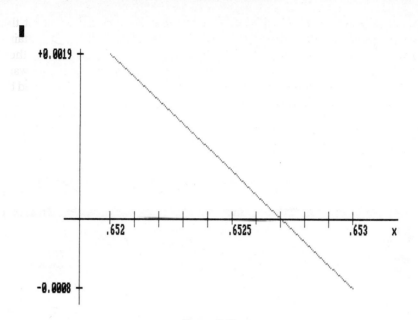

Figure 1.42

how this computational power can be used to illuminate some of the basic concepts of calculus.

PROBLEMS

1. Use Program ZOOMSOLV to find the other two solutions of the cork ball equation, with four-place accuracy.
2-6. Select five other equations mentioned in this chapter, and use Program ZOOMSOLV to find their solutions visually.
7. Write a program to plot simultaneously the graphs of two given functions $f(x)$ and $g(x)$ on an input interval $[a, b]$ so that their points of intersection can be spotted. You will need to select a scale for the y-axis that "works" with both functions.

Limits and Derivatives

2

2.1 SEQUENCES AND THE LIMIT CONCEPT

In Chapter 1 we discussed several iterative methods for producing a *sequence* $x_1, x_2, x_3, \ldots$ of closer and closer approximations to a solution r of an equation $f(x) = 0$. Now we need to examine precisely what it means to say that the approximations $\{x_n\}$ actually "approach" the solution r as n increases.

This involves the concept of the *limit* of a sequence $\{x_n\}$ of numbers $x_1, x_2, x_3, \ldots$. We say that this sequence of numbers *converges* to the number r provided that we can make the number x_n as close as we please to the number r by choosing the index n sufficiently large. In this case we also say that r is the *limit* of the sequence $\{x_n\}$, and write

$$\lim_{n \to \infty} x_n = r. \tag{1}$$

We may also write $x_n \longrightarrow r$ as $n \longrightarrow \infty$. In terms of decimal expansions, it is equivalent to say that for any positive integer k, the numbers x_n and r agree to k decimal places when n is sufficiently large.

For instance, suppose the sequence $\{x_n\}$ is defined by

$$x_1 = 0.3, \quad x_2 = 0.33, \quad x_3 = 0.333, \quad \ldots.$$

Then the limit of this sequence is

$$r = \frac{1}{3} = 0.3333 \ldots,$$

because x_n and r agree to n decimal places. Similarly, if

$$x_1 = 0.27, \quad x_2 = 0.2727, \quad x_3 = 0.272727, \quad \ldots,$$

then x_n approaches the number

$$\frac{3}{11} = 0.272727 \ldots$$

as $n \longrightarrow \infty$. The sequence $\{1/n\}$ obviously converges to 0; in particular, the numbers $x_n = 1/n$ and 0 agree to k decimal places whenever $n \geq 10^{k+1}$. For instance, if $n \geq 10^4 = 10{,}000$, then $0 < 1/n \leq 10^{-4} = 0.0001$, so the number $x_n = 1/n$ has a zero in each of its first three decimal places.

If x_n fails to approach any number at all as $n \longrightarrow \infty$, then the limit of the sequence $\{x_n\}$ does not exist, and we say that the sequence *diverges*. There are two principal ways in which a sequence can diverge. One way is illustrated by the sequence $\{n^2\}$ of squares of the positive integers. In this case, $x_n = n^2$ increases without bound as $n \longrightarrow \infty$, and we may say that the sequence *diverges to infinity*. By contrast, if $x_n = 1 + (-1)^n$, then $x_n = 0$ if n is odd, while $x_n = 2$ if n is even. Thus x_n oscillates between the values 0 and 2, and we may say that the sequence $\{1 + (-1)^n\}$ *diverges by oscillation*.

Much more complicated modes of divergence are possible. For example, suppose that

$$x_n = \frac{(-1)^n}{n} + n \bmod 5,$$

where $n \bmod p$ denotes the remainder when n is divided by p; for instance, $17 \bmod 5 = 2$. Then x_n cycles periodically through neighborhoods of the distinct numbers 0, 1, 2, 3, and 4.

Compound Interest and the Number e

The problem of calculating compound interest leads to an important and interesting sequence. Suppose we deposit an amount of A_0 dollars in a savings account that accrues interest at an annual interest rate r; for instance, $r = 0.08$ corresponds to an 8 percent annual interest rate. If *simple interest* is computed after T years and added to the account, then the new amount in the account is

$$A = A_0 + A_0 rT = A_0(1 + rT). \tag{2}$$

Thus the accrual of simple interest for time T multiplies the original amount by the factor $(1 + rT)$. For instance, if $A_0 = \$1$, $r = 0.08$, and $T = 1$ year, then (2) gives $1.08, so each original dollar "grows" in 1 year to $1.08 at 8 percent annual interest.

Compound interest means the result of accruing simple interest regularly at specified intervals of time. To compound interest *quarterly*, for instance, we let $T = 1/4$ year in (2) and multiply the amount in the account by $(1 + r/4)$ at the end of each quarter. In 1 year we do this four times, so the amount in the account after 1 year is

$$A_4 = A_0\left(1 + \frac{r}{4}\right)^4. \tag{3}$$

With $A_0 = \$1$ and $r = 0.08$, this formula yields $A_4 = 1.0824$. Thus a deposit of \$100 will earn \$8.24 in 1 year if 8 percent annual interest is compounded quarterly. This is what a bank means when it advertises that it increases the "effective annual interest rate" from 8 percent to 8.24 percent by compounding quarterly.

Let us now take $T = 1/n$ in (2), so that the interest is compounded n times per year. The effect in 1 year is to multiply the amount n times by the factor $(1 + r/n)$. Hence the amount in the account after 1 year is

$$A_n = A_0\left(1 + \frac{r}{n}\right)^n. \tag{4}$$

Figure 2.1 shows the values of A_n, with $A_0 = 1$ and $r = 0.08$, for some common values of n. For instance, we see that the result of compounding 8 percent interest daily is an effective annual interest rate of 8.33 percent.

n	A_n	$A_n - A_0$	
1	1.0800	0.0800	(annually)
4	1.0824	0.0824	(quarterly)
12	1.0830	0.0830	(monthly)
52	1.0832	0.0832	(weekly)
365	1.0833	0.0833	(daily)

Figure 2.1 Compounding 8 percent interest n times during 1 year

To get the sequence we are really interested in, we let $A_0 = r = 1$ in (4) and write x_n instead of A_n. Thus the sequence $\{x_n\}$ is defined by

$$x_n = \left(1 + \frac{1}{n}\right)^n \tag{5}$$

for $n = 1, 2, 3, \ldots$, and x_n may be regarded as the yield in 1 year when \$1 is invested at 100 percent interest compounded n times annually. The data in Figure 2.1 suggest that the value of x_n will increase as n increases. Long ago the question arose as to whether there is an upper bound to the value of x_n.

```
100 REM--Program LIMSEQ (preliminary version)
110 REM--Computes the limit (if any) of the sequence
120 REM--whose nth term is defined in line 180, with
130 REM--the error tolerance specified in line 200.
140 REM
150       N = 0 : Y = 0
160       FOR I = 1 TO 10
170           N = N + 1
180           X = (1 + 1/N)^N
190           PRINT N, X
200           IF ABS(X - Y) < .000001 THEN GOTO 260
210           Y = X
220       NEXT I
230       INPUT "WANT 10 MORE VALUES (Y/N)"; A$
240       IF A$ = "Y" OR A$ = "y" THEN GOTO 160
250       STOP
260       PRINT "LIMIT = "; X
270       END
```

```
RUN
  1                2
  2                2.25
  3                2.370371
  4                2.441406
  5                2.48832
  6                2.521626
  7                2.546499
  8                2.565785
  9                2.581176
 10                2.593743
WANT 10 MORE VALUES (Y/N)? Y


 91                2.703512
 92                2.703669
 93                2.703805
 94                2.703971
 95                2.704098
 96                2.704245
 97                2.704391
 98                2.704536
 99                2.704666
100                2.704813
WANT 10 MORE VALUES (Y/N)?
```

Figure 2.2   Preliminary version of Program LIMSEQ

Can the effective annual interest be increased indefinitely by compounding more and more frequently? In mathematical terms, the question is whether the sequence $\{x_n\}$ defined by (5) has a (finite) limit.

Figure 2.2 shows the result of a first attempt to write a program to investigate this question. The program shown simply calculates and displays the values $x_1, x_2, x_3, \ldots$ in blocks of 10. Note that Formula (5) defining the sequence $\{x_n\}$ appears in line 180. On each pass through the FOR-NEXT loop of lines 160–220, X denotes the new value x_n and Y denotes the previous value x_{n-1}. Unless the condition $|x_n - x_{n-1}| < 0.000001$ is satisfied in line 200, the loop is completed and line 230 asks whether we want 10 more values calculated.

The results included in Figure 2.2 show that after calculating the first 100 values we are still nowhere close to five-place accuracy. They may even suggest that the limit is approximately 2.704, but actually its value is 2.718 to three decimal places. The problem is that we are not moving rapidly enough through the values in the sequence. How can we speed up the process?

```
100 REM--Program LIMSEQ (second preliminary version)
110 REM--Computes the limit (if any) of the sequence
120 REM--whose nth term is defined in line 180, with
130 REM--the error tolerance specified in line 200.
140 REM
150       N = 1 : Y = 0
160       FOR I = 1 TO 10
170           N = 10*N
180           X = (1 + 1/N)^N
190           PRINT N, X
200           IF ABS(X - Y) < .000001 THEN GOTO 260
210           Y = X
220       NEXT I
230       INPUT "WANT 10 MORE VALUES (Y/N)"; A$
240       IF A$ = "Y" OR A$ = "y" THEN GOTO 160
250       STOP
260       PRINT "LIMIT = "; X
270       END
```

```
RUN
  10              2.593743
  100             2.704813
  1000            2.717042
  10000           2.718023
  100000          2.695423
  1000000         2.481655
  1E+07           2.28484
  1E+08           1
  1E+09           1
LIMIT =   1
```

Figure 2.3   Another version of Program LIMSEQ

If, indeed, the sequence $\{x_n\}$ has a limit, it should be apparent that the *subsequence*

$$x_{10}, \quad x_{100}, \quad x_{1000}, \quad x_{10,000}, \quad \ldots \tag{6}$$

has the same limit. The sequence in (6) consists only of those elements of the original sequence whose subscripts are powers of 10. Figure 2.3 shows a slight alteration—see line 170—of our program that picks out for computation just those values that appear in the subsequence (6).

But look at the RUN shown in Figure 2.3. The values of X increase at first (as they should), reaching 2.718 when $n = 10,000$. But then they start decreasing, and finally collapse altogether; the program prints out the false result LIMIT $= 1$. Now what's wrong?

What we are seeing here is a drastic case of *round-off error*. In single-precision BASIC the value 1.00000001 is rounded off to 1, and this is why we get $x = 1$ when $n \geq 10^8$.

```
100 REM--Program LIMSEQ
110 REM--Computes the limit (if any) of the sequence
120 REM--whose nth term is defined in line 200, with
130 REM--the error tolerance specified in line 220.
140 REM
150     DEFDBL  X, N
160     PRINT "  N", "   X"  :   PRINT
170     N = 1 : XOLD = 0
180     FOR I=1 TO 10
190         N = 10*N
200         XNEW = (1 + 1/N)^N
210         PRINT N,  :  PRINT USING "#.#####"; XNEW
220         IF ABS(XNEW - XOLD) < .000001 THEN GOTO 280
230         XOLD = XNEW
240     NEXT I
250     INPUT "WANT 10 MORE VALUES (Y/N)"; A$
260     IF A$ = "Y" OR A$ = "y" THEN GOTO 180
270     STOP
280     PRINT
290     PRINT USING "LIMIT = #.#####"; XNEW
300     END
```

Listing 2.4   Program LIMSEQ

Listing 2.4 shows our final improved version of Program LIMSEQ. We have attempted to "cure" the round-off problem with the double-precision declaration in line 150. Even so, there is a hidden difficulty. The quantity $(1 + 1/n)^n$ is actually computed in line 200 as

$$X = EXP(N * LOG(1 + 1/N)),$$

where EXP(X) denotes the exponential function e^x and LOG(X) denotes the natural logarithm $\ln x$ in BASIC. The difficulty is that in "ordinary" BASIC, the exponential and logarithm functions are calculated only in single precision (even when all variables involved have been declared double precision).

Figure 2.5 shows a run of Program LIMSEQ under version 2.00 of IBM BASIC using the BASICA/D option, which calculates exponential and logarithmic values in double precision. We finally get LIMIT $= 2.71828$, which is the correct value to five decimal places. The number

$$e = \lim_{n \to \infty} \left(1 + \frac{1}{n}\right)^n \tag{7}$$

```
RUN
  N                    X

  10             2.59374
  100            2.70481
  1000           2.71692
  10000          2.71815
  100000         2.71827
  1000000        2.71828
  10000000       2.71828
  100000000      2.71828

  LIMIT = 2.71828
```

Figure 2.5 Run of Program LIMSEQ to compute the number e

is the most important special number in higher mathematics. As we will see in Chapter 4, its value to 15 decimal places is

$$e = 2.718281828459045 \dots . \tag{8}$$

Despite the possibility of periodicity that seems apparent in (8), the number e is irrational, so its infinite decimal expansion is *not* periodic. A final remark: Even with all the double precision described above, Program LIMSEQ is very susceptible to round-off errors with this particular sequence, and $e \approx 2.71828$ is about as much accuracy as can be wrung out of it. Try running the program with a smaller error tolerance in line 220 but without PRINT USING to see what happens.

Program LIMSEQ can be applied to investigate any sequence $\{x_n\}$ after editing the formula for x_n (in terms of n) into line 200. However, you should now realize that merely because the program prints out LIMIT = whatever, this does not necessarily mean that it's so. In addition to the possibility of round-off errors, it can happen that the subsequence defined in (6) converges, whereas the full sequence $\{x_n\}$ does not. For example, suppose that

$$x_n = \sin \frac{n\pi}{10} \tag{9}$$

for each $n = 1, 2, 3, \dots$. Then LIMSEQ indicates that the limit is 0, because

$$\sin \frac{10^n\pi}{10} = \sin 10^{n-1}\pi = 0.$$

However, there are arbitrarily large integers of the form $n = 20k + 5$ such that

$$\sin \frac{n\pi}{10} = \sin \frac{(20k+5)\pi}{10} = \sin \left(2k + \frac{1}{2}\right)\pi$$

$$= \sin \frac{\pi}{2} = 1.$$

Thus the sequence $\{\sin n\pi/10\}$ does not converge; Program LIMSEQ gives a false result because it tests only a particular subsequence that happens to converge.

Another potential difficulty is illustrated by the sequence $\{x_n\}$ defined by

$$x_n = 1 + \frac{1}{2} + \frac{1}{3} + \cdots + \frac{1}{n}, \tag{10}$$

the sum of the reciprocals of the first n positive integers. In Section 12.3 of Edwards and Penney, *Calculus* it is proved that the sequence defined in (10) *diverges* to infinity. That is, however large M may be, it is eventually true that $x_n > M$. Nevertheless, Program LIMSEQ will print out an alleged limit as soon as $1/n$ is less than the error tolerance used, because

$$x_n - x_{n-1} = \frac{1}{n}.$$

When, then, can Program LIMSEQ be trusted? The simplest answer is that no single computer program can handle infallibly all possible sequences, but LIMSEQ is reasonably reliable when it is known in advance that the given sequence converges, for then every subsequence must approach the same limit. For example, consider the sequence $\{x_n\}$ defined by $x_n = \sqrt[n]{2}$. It should be apparent that (a) $x_n > x_{n+1}$ and (b) $1 \leq x_n \leq 2$ for every n. Thus $\{x_n\}$ is a bounded monotone decreasing sequence, and every such sequence is known to converge. We could therefore use LIMSEQ confidently to find the limit of $\sqrt[n]{2}$ as $n \longrightarrow \infty$.

PROBLEMS

Use Program LIMSEQ (editing line 200 in each case) to investigate the limits in Problems 1–8.

1. $\lim\limits_{n\to\infty} \sqrt[n]{2} = 1$

2. $\lim\limits_{n\to\infty} \sqrt[n]{n} = 1$

3. $\lim\limits_{n\to\infty} \dfrac{1}{n} \sin \dfrac{n\pi}{10} = 0$

4. $\lim\limits_{n\to\infty} n \sin \dfrac{1}{n} = 1$

5. $\lim\limits_{n\to\infty} (\sqrt{n + 10} - n) = 0$

6. $\lim\limits_{n\to\infty} (\sqrt{n^2 + 10} - n) = 0$

7. $\lim\limits_{n\to\infty} \dfrac{n - 1}{2n + 1} = \dfrac{1}{2}$

8. $\displaystyle\lim_{n\to\infty} \frac{1^2 + 2^2 + \cdots + n^2}{n^3} = \frac{1}{3}$

9. It is known that the natural logarithm is given by

$$\ln x = \lim_{n\to\infty} n(\sqrt[n]{x} - 1).$$

Use Program LIMSEQ to check the fact that $\ln 2 \approx 0.69315$.

10. Euler's constant γ is defined by

$$\gamma = \lim_{n\to\infty}\left(1 + \frac{1}{2} + \frac{1}{3} + \cdots + \frac{1}{n} - \ln n\right).$$

Write a program to check the fact that $\gamma \approx 0.57722$. Print only every tenth value, and stop when $|x_n - 0.577| < 0.0005$. Be prepared to wait awhile.

11. Let the two sequences $\{a_n\}$ and $\{b_n\}$ be defined by $a_0 = 2$, $b_0 = 3$, and

$$a_{n+1} = \frac{a_n + b_n}{2}, \qquad b_{n+1} = \sqrt{a_n b_n}.$$

Write a program to check the fact that both sequences converge to the same limit.

12. According to *Stirling's formula*, the factorial

$$n! = n(n - 1)(n - 2) \cdots 3 \cdot 2 \cdot 1$$

is given approximately by

$$n! \approx \sqrt{2\pi n}\left(\frac{n}{e}\right)^n;$$

more precisely,

$$\lim_{n\to\infty} \frac{n!}{\sqrt{2\pi n}\,(n/e)^n} = 1.$$

Use Program LIMSEQ to test this limit. Also, write a short program to print values of n, $n!$, and $\sqrt{2\pi n}\,(n/e)^n$ for comparison.

13. Leonardo Fibonacci asked (in his *Liber Abaci* published in 1202) how many pairs of rabbits can be produced from a single pair in a year if every month each pair begets a new pair which is similarly productive beginning in the second succeeding month. The answer is provided by the *Fibonacci sequence* $\{F_n\}$, which is defined by

$$F_1 = 1, \quad F_2 = 1, \quad F_{n+1} = F_n + F_{n-1} \qquad \text{for } n \geqq 2.$$

Write a program that computes the numbers $F_1, F_2, F_3, \ldots$ and also verifies that

$$\lim_{n \to \infty} \frac{F_{n+1}}{F_n} = \frac{1}{2}(1 + \sqrt{5}).$$

2.2 LIMITS OF FUNCTIONS

The limits of sequences that we discussed in Section 2.1 are closely related to limits of functions. Suppose the function $f(x)$ is defined on an open interval containing the point $x = a$, except possibly at the point a itself. Then we can talk about the value of $f(x)$ when x is close to a, and ask what happens to $f(x)$ as x gets closer and closer to a, that is, as x *approaches a*.

For example, suppose that

$$f(x) = \frac{x - 1}{x + 2} \tag{1}$$

and $a = 3$. Then when x is close to 3, the number $x - 1$ is close to 2 and the denominator $x + 2$ is close to 5, so the value of the fraction $f(x)$ ought to be close to the quotient $2/5$. That is, it appears that the value $f(x)$ approaches $2/5$ as x approaches 3: $f(x) \longrightarrow 2/5$ as $x \longrightarrow 3$. Assuming that this is actually so, we write

$$\lim_{x \to 3} \frac{x - 1}{x + 2} = \frac{2}{5} \tag{2}$$

and say that the limit as x approaches 3 of $f(x) = (x - 1)/(x + 2)$ is $2/5$.

Figure 2.6 shows a simple program written to investigate the limit in (2). It computes the values of $f(x)$ for the points $3.1, 3.01, 3.001, \ldots$ approaching 3 from the right, and also for the points $2.9, 2.99, 2.999, \ldots$ approaching 3 from the left. The results of the run shown in the figure certainly reinforce our guess that $f(x)$ approaches $2/5 = 0.4$ as x approaches 3.

We say (by definition) that the *limit* of the function $f(x)$ is L as x approaches a, and write

$$\lim_{x \to a} f(x) = L, \tag{3}$$

provided that the value $f(x)$ can be made as close to L as one pleases by choosing x sufficiently near (although not equal) to the number a. What this means, roughly, is that $f(x)$ gets very close to L when x gets very close to a.

```
100 REM--Program PRELIM1
110 REM--Preliminary version of Program LIMFCTN
120 REM
130 DEF FNF(X) = (X - 1)/(X + 2)
140 PRINT " x", "f(x)", "x", "f(x)"
150 PRINT
160 A = 3 : H = .1
170 FOR I = 1 TO 6
180     PRINT A + H,
190     PRINT USING "#.#####       "; FNF(A + H),
200     PRINT A - H,
210     PRINT USING "#.#####"; FNF(A - H)
220     H = H/10
230 NEXT I
240 END
```

```
RUN
x                 f(x)              x              f(x)

3.1               0.41176           2.9            0.38776
3.01              0.40120           2.99           0.39880
3.001             0.40012           2.999          0.39988
3.0001            0.40001           2.9999         0.39999
3.00001           0.40000           2.99999        0.40000
3.000001          0.40000           2.999999       0.40000
```

Figure 2.6   Investigating the limit in Equation (2)

In particular, in order that $f(x) \longrightarrow L$ as $x \longrightarrow a$, it must be true that $f(x_n) \longrightarrow L$ for every sequence $\{x_n\}$ of points (no one of them equal to a) that converge to a. Although it is not so obvious, the converse of this statement is also true. Hence the connection between limits of sequences and limits of functions is this:

$$\lim_{x \to a} f(x) = L \tag{3}$$

if and only if

$$\lim_{n \to \infty} f(x_n) = L \tag{4}$$

for *every* such sequence $\{x_n\}$ of points approaching a.

We therefore can investigate the functional limit in (3) by investigating—as in Section 2.1—the sequential limit in (4). However, no computer program can check Condition (4) simultaneously for *all* possible sequences $\{x_n\}$ converging to a. As a practical matter, the most we can expect is to verify (4) for one or two "typical" sequences $\{x_n\}$ of points approaching a. When we find that $f(x_n) \longrightarrow L$ for a *particular* sequence of points $\{x_n\}$ approaching a, this suggests—without proving it conclusively—that $f(x) \longrightarrow L$ as $x \longrightarrow a$. But there remains the possibility that for some *other* such sequence $\{x_n\}$, $\{f(x_n)\}$ does *not* approach L as $n \longrightarrow \infty$.

For instance, in the case of the function $f(x) = (x - 1)/(x + 2)$ in (2), we see from the data in Figure 2.6 that $f(x_n) \longrightarrow 0.4$ as $n \longrightarrow \infty$, when either

$$x_n = 3 + \frac{1}{10^n} \qquad \text{or} \qquad x_n = 3 - \frac{1}{10^n}.$$

This makes it seem almost certain that $f(x) \longrightarrow 0.4$ as $x \longrightarrow 3$, and indeed this is true.

For a more deceptive example, consider the function

$$f(x) = \sin \frac{\pi}{x}. \tag{5}$$

We want to investigate the limit of $f(x)$ as $x \longrightarrow 0$. The sequence $\{1/n\}$ with $x_n = 1/n$ converges to 0, and

$$f(x_n) = f\left(\frac{1}{n}\right) = \sin \frac{\pi}{1/n} = \sin n\pi = 0$$

for all n, so certainly $f(x_n) \longrightarrow 0$ as $n \longrightarrow \infty$. However, it would be premature to conclude that $f(x) \longrightarrow 0$ as $x \longrightarrow 0$. To see why, consider the second sequence $\{2/(4n + 1)\}$ of points converging to 0. Now

$$f(x_n) = f\left(\frac{2}{4n + 1}\right) = \sin (4n + 1)\frac{\pi}{2}$$

$$= \sin \left(2n\pi + \frac{\pi}{2}\right) = \sin \frac{\pi}{2} = 1$$

for all n. So the limit is not 0, after all. What happens here is that as $x \longrightarrow 0$, $f(x) = \sin \pi/x$ oscillates repeatedly between the values $+1$ and -1, and the limit of $f(x)$ as $x \longrightarrow 0$ *does not exist*.

It is when we have reason to suspect that the value of $f(x)$ "oscillates" near $x = a$ that we need to be especially cautious. Otherwise, we are reasonably safe in concluding that $f(x) \longrightarrow L$ as $x \longrightarrow a$ if we can verify that $f(x_n) \longrightarrow L$ for a single "typical" sequence $\{x_n\}$ of points approaching a.

The following example illustrates the fact that the value of the function f at the point $x = a$ itself is irrelevant, and need not even be defined.

Example 1

Investigate the limit

$$\lim_{x \to 2} \frac{(x + 1)^2 - 9}{x - 2}. \tag{6}$$

```
100 REM--Program PRELIM2
110 REM--Preliminary version of Program LIMFCTN
120 REM
130 DEFDBL X, F, A, H
140 DEF FNF(X) = ((X + 1)*(X + 1) - 9)/(X - 2)
150 PRINT " x                      f(x)"
160 PRINT
170 A = 2 : H = .1
180 FOR I = 1 TO 7
190     X = A + H
200     PRINT USING "#.#######"; X,
210     PRINT USING "            #.#####"; FNF(X)
220     H = H/10
230 NEXT I
240 END
```

```
RUN
 x                      f(x)

2.1000000            6.10000
2.0100000            6.01000
2.0010000            6.00100
2.0001000            6.00010
2.0000100            6.00001
2.0000010            6.00000
2.0000001            6.00000
```

Figure 2.7   Investigating the limit in Equation (6)

Solution   We certainly cannot simply substitute the value $x = 2$, because then $f(x) = [(x + 1)^2 - 9]/(x - 2)$ would reduce to the meaningless form $0/0$. However, the results shown in Figure 2.7 convince us that $f(x)$ approaches the number 6 as x approaches 2. In this case the apparent value of the limit can be verified by the following algebraic manipulation:

$$\lim_{x \to 2} \frac{(x + 1)^2 - 9}{x - 2} = \lim_{x \to 2} \frac{x^2 + 2x - 8}{x - 2}$$

$$= \lim_{x \to 2} \frac{(x + 4)(x - 2)}{x - 2}$$

$$= \lim_{x \to 2} (x + 4) = 6.$$

Program LIMFCTN in Listing 2.8 is a general program for investigating numerically the limit of a function $f(x)$ as x approaches a. We have included a few extra bells and whistles. Note that the function $f(x)$ and the number a are left blank in lines 280 and 290. When the program is run, it will pause (in response to lines 240 and 260) for you to type in and enter (by carriage return) the particular function $f(x)$ you want, as well as the point a where you want its limit. Line 320 permits us to choose whether we want to use a sequence $\{x_n\}$ of points approaching a from the right or from

```
100 REM--Program LIMFCTN
110 REM--Investigates the limit of a function f(x) as x
120 REM--approaches a by examining the values {f(Xn)} for
130 REM--a random sequence {Xn} of points approaching a.
135 REM--Line 300 requests that a random "seed" be entered.
140 REM--The program pauses for the function f(x) to be
150 REM--edited into line 280.  After doing so, continue by
160 REM--entering the command  RUN 250.  Similiarly, the
170 REM--value  x = a  is to be edited into line 290.
180 REM--Finally line 310 asks whether you want a lefthand
190 REM--limit or a righthand limit.
200 REM
210 REM--Initialization:
220 REM
230        PRINT "EDIT IN YOUR FUNCTION F(X), THEN RUN 250":
           PRINT
240        EDIT 280
250        PRINT "EDIT IN YOUR POINT A, THEN RUN 270": PRINT
260        EDIT 290
270        DEFDBL F, X, A, H
280        DEF FNF(X) =
290        A =
300        RANDOMIZE    :    PRINT
310        INPUT "WANT LEFTHAND OR RIGHTHAND LIMIT (L/R)"; C$
320        IF C$ = "L" OR C$ = "l" THEN LET H = -1 ELSE H = 1
330        PRINT " x                   f(x)"
340        PRINT
350 REM
360 REM--Compute sequence values:
370 REM
380        FOR I = 1 TO 7
390            X = A + H*RND
400            PRINT USING "+#.#######"; X,
410            PRINT USING "          +#.#######"; FNF(X)
420            H = H/10
430        NEXT I
440        END
```

Listing 2.8   Program LIMFCTN

the left—for the limit to exist, the same limiting value must be approached from both sides (see Problem 12). Finally, because of lines 300 and 390, we employ an essentially *random* sequence of points $\{x_n\}$ approaching a, with

$$x_n = a + \frac{\text{RND}(n)}{10^n}, \tag{7}$$

where $\text{RND}(n)$ denotes different "random numbers" between 0 and 1 (if the right-hand limit is chosen) for different values of n. With different random number seeds input at line 300 we get different sequences of random numbers $\{\text{RND}(n)\}$ in Equation (7). If we get the same apparent limit in multiple runs, from both sides and with different seeds, we can be rather confident of its value.

Example 2

Verify the limit

$$\lim_{x \to 0} \frac{\sin x}{x} = 1, \tag{8}$$

which is important in the calculus of trigonometric functions. Here sin x denotes the sine of an angle of x *radians*—SIN(X) in BASIC.

```
RUN
EDIT IN YOUR FUNCTION F(X), THEN RUN 250

280      DEF FNF(X) =   (SIN(X))/X
RUN 250
EDIT IN YOUR POINT A, THEN RUN 270

290      A = 0
RUN 270
Random number seed (-32768 to 32767)? 3

WANT LEFTHAND OR RIGHTHAND LIMIT (L/R)? R
x                      f(x)

+0.8671320            +0.8793084
+0.0918361            +0.9985950
+0.0069064            +0.9999920
+0.0005842            +1.0000000
+0.0000319            +1.0000000
+0.0000081            +1.0000000
+0.0000003            +1.0000000
```

Figure 2.9 A run of Program LIMFCTN with $f(x) = (\sin x)/x$

```
RUN
EDIT IN YOUR FUNCTION F(X), THEN RUN 250

280      DEF FNF(X) =   (SIN(X))/X
RUN 250
EDIT IN YOUR POINT A, THEN RUN 270

290      A = 0
RUN 270
Random number seed (-32768 to 32767)? 17

WANT LEFTHAND OR RIGHTHAND LIMIT (L/R)? L
x                      f(x)

-0.7128046            +0.9174438
-0.0393062            +0.9997425
-0.0027825            +0.9999987
-0.0008096            +1.0000000
-0.0000666            +1.0000000
-0.0000078            +1.0000000
-0.0000004            +1.0000000
```

Figure 2.10 Another run of Program LIMFCTN with $f(x) = (\sin x)/x$

Solution. Figures 2.9 and 2.10 show the results of two runs of Program LIMFCTN with $f(x) = (\sin x)/x$ and $a = 0$. Note the different "random" sequences of values of x approaching 0 from the right and from the left, respectively. The results leave little doubt that the value of the limit in (8) is 1. This fact is proved analytically in Section 2.5 of Edwards and Penney, *Calculus*.

Remark: Despite the double-precision declaration in line 270, only single-precision results are shown in Figures 2.9 and 2.10. Whenever the function $f(x)$ involves roots or powers, exponentials, logarithms, or trigonometric functions, Program LIMFCTN must be run in version 2.0 of IBM BASIC in order to actually get double-precision results.

Example 3

Investigate the limit

$$\lim_{x \to 0} \frac{1}{x} \ln \frac{7x + 8}{4x + 8}.$$

(9)

Here $\ln x$ denotes the natural (base e) logarithm function—LOG(X) in BASIC.

```
RUN
EDIT IN YOUR FUNCTION F(X), THEN RUN 250

280      DEF FNF(X) =  LOG((7*X+8)/(4*X+8))/X
RUN 250
EDIT IN YOUR POINT A, THEN RUN 270

290      A = 0
RUN 270
Random number seed (-32768 to 32767)? 5

WANT LEFTHAND OR RIGHTHAND LIMIT (L/R)? R
x                       f(x)

+0.2736566663          +0.3158161
+0.0710505486          +0.3575546
+0.0033764213          +0.3741316
+0.0006197454          +0.3748403
+0.0000224302          +0.3749942
+0.0000009319          +0.3749998
+0.0000004271          +0.3749999
+0.0000000603          +0.3750000
+0.0000000053          +0.3750000
+0.0000000004          +0.3750000
```

Figure 2.11 Investigating the limit of Example 3

Solution. An initial run of Program LIMFCTN in version 1.1 BASIC gave inconclusive results because of round-off error. Figure 2.11 shows the results of a run in version 2.0 BASIC with the double-precision option. Evidently, the value of the limit in (9) is $0.375 = 3/8$, and this can be verified using l'Hospital's rule of calculus (see Section 11.1 of Edwards and Penney, *Calculus*).

x	f(x)
0.1	0.97427
0.01	0.95340
0.001	0.94139
0.0001	0.93297
0.00001	0.92648
10^{-10}	0.90663
10^{-20}	0.88720
10^{-40}	0.86819
10^{-60}	0.85726
10^{-80}	0.84959
10^{-100}	0.84369

Figure 2.12   The Function $f(x)$ in Equation (10)

A final caveat: Some limits are approached so slowly that practical numerical computations using LIMFCTN can never reveal their values. For instance, let

$$f(x) = \left(\ln \frac{1}{x} \right)^{-32}. \tag{10}$$

Then $\ln (1/x) = -\ln x \longrightarrow \infty$ as $x \longrightarrow 0$, so it follows that $f(x) \longrightarrow 0$ as $x \longrightarrow 0$. But look at the values of $f(x)$ shown in Figure 2.12. They would hardly suggest to anyone that the limit of $f(x)$ is zero as $x \longrightarrow 0$.

PROBLEMS

Investigate numerically the limits in Problems 1-11.

1. $\lim\limits_{x \to 0} \dfrac{(1 + x)^2 - 1}{x}$

2. $\lim\limits_{x \to 2} \dfrac{x^3 - 8}{x - 2}$

3. $\lim\limits_{x \to 1} \dfrac{x^3 - 1}{x^2 + 4x - 5}$

4. $\lim\limits_{x \to 11} \dfrac{(x - 10)^2 - 1}{x - 11}$

5. $\lim\limits_{x \to 1} \dfrac{x^5 - 1}{x - 1}$

6. $\lim\limits_{x \to 4} \dfrac{\sqrt{x} - 2}{x - 4}$

7. $\lim\limits_{x \to 0} \dfrac{1 - \cos x}{x^2}$

8. $\lim\limits_{x \to 0} \dfrac{x - \tan x}{x^3}$

9. $\lim\limits_{x \to 0} \dfrac{1}{x} \ln \dfrac{1-x}{1+x}$

10. $\lim\limits_{x \to 0} \left(\dfrac{1}{x}\right)^{1/x}$

11. $\lim\limits_{x \to 0} (1+x)^{1/x}$

12. Show that the right-hand and left-hand limits of $f(x) = 1/(1 + e^{1/x})$ as $x \to 0$ are unequal.

Verify numerically the limits in Problems 13 and 14.

13. $\lim\limits_{x \to 0} \left(1 + \dfrac{x}{2}\right)^{1/x} = \sqrt{e}$

14. $\lim\limits_{x \to 0} (\cos x)^{1/x^2} = \dfrac{1}{\sqrt{e}}$

2.3 NUMERICAL DIFFERENTIATION

The concept of differentiation in calculus stems from the problem of finding the *tangent line* at the point $P(a, f(a))$ to the curve $y = f(x)$ in the xy- plane. The key idea is to consider a nearby second point $Q(a + h, f(a + h))$ on the curve (see Figure 2.13), where $h \neq 0$ denotes the difference in the x-coordinates of P and Q. The slope of the *secant line* through P and Q—the difference of the y-coordinates divided by the difference of the x-coordinates—is

$$m_{\text{sec}} = \frac{f(a+h) - f(a)}{(a+h) - a} = \frac{f(a+h) - f(a)}{h}. \tag{1}$$

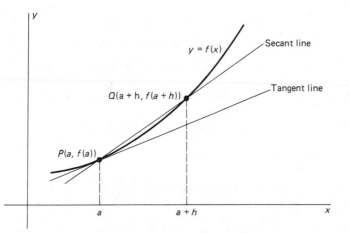

Figure 2.13 The tangent line and the secant line

It appears intuitively that if the number h is sufficiently small in magnitude, then this secant line should lie very close to the tangent line through P. We therefore *define* the tangent line to be the line through the point P whose slope m is given by

$$m = \lim_{h \to 0} m_{\text{sec}} = \lim_{h \to 0} \frac{f(a + h) - f(a)}{h}. \tag{2}$$

Of course, the slope m depends on the x-coordinate a of the point P. If we now let a vary, we get a new function $f'(x)$ defined by

$$f'(x) = \lim_{h \to 0} \frac{f(x + h) - f(x)}{h}. \tag{3}$$

This new function $f'(x)$ is called the *derivative* of the original function $f(x)$, and the process of finding $f'(x)$ explicitly is called *differentiation* of $f(x)$. The following table lists some of the derivatives of familiar elementary functions that one learns in an introductory calculus course. For instance, if $f(x) = x^2$, then $f'(x) = 2x$.

Function $f(x)$	Derivative $f'(x)$
1	0
x^k	kx^{k-1}
$\sin kx$	$k \cos kx$
$\cos kx$	$-k \sin kx$
$\ln x$	$\dfrac{1}{x}$
e^{kx}	ke^{kx}

If $f(x) = \sin 3x$, then $f'(x) = 3 \cos 3x$. If $f(x) = e^{4x}$, then $f'(x) = 4e^{4x}$.

Moreover, the operation of differentiation is *linear*, meaning that if

$$h(x) = af(x) + bg(x),$$

where a and b are constants, then

$$h'(x) = af'(x) + bg'(x).$$

For instance, if

$$p(x) = 2x^3 - 3x^2 + 4x - 5,$$

then

$$p'(x) = 2(3x^2) - 3(2x) + 4(1) - 5(0)$$
$$= 6x^2 - 6x + 4.$$

Thus the term "derivative" here means the same thing for polynomials as it meant in Section 1.5.

However, our interest in this section lies not in differentiation of functions using the formulas above, but rather in the numerical evaluation of the limit in (3) that defines the derivative. This is *numerical differentiation.* Let us write

$$f'(x_0) = \lim_{h \to 0} \frac{f(x_0 + h) - f(x_0)}{h} \tag{4}$$

to emphasize the specific fixed point x_0 at which we want to compute the numerical value of the derivative of the function f. Note that

$$f'(x_0) = \lim_{h \to 0} D(h) = \lim_{x \to 0} D(x) \tag{5}$$

where

$$D(h) = \frac{f(x_0 + h) - f(x_0)}{h} \tag{6}$$

is the *difference quotient* of the function f at the point x_0. One way to compute $f'(x_0)$ is to run Program LIMFCTN (Listing 2.8) with $a = 0$ and with FNF(X) defined in line 280 as $D(x)$.

Example 1

Find the slope m of the tangent line to the circle $x^2 + y^2 = 25$ at the point $(3, 4)$.

Solution.   From the right triangle in Figure 2.14, together with the fact that the slopes of perpendicular lines are negative reciprocals of each other, we find that $m = -3/4$. We want to verify this value numerically. The upper semicircle shown is the graph of the function

$$y = f(x) = \sqrt{25 - x^2}.$$

We have $x_0 = 3$ and $f(x_0) = 4$, so the difference quotient in (6) is

$$D(h) = \frac{\sqrt{25 - (3 + h)^2} - 4}{h}. \tag{7}$$

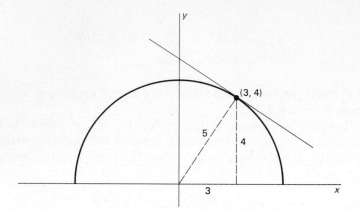

Figure 2.14   Tangent line to the circle $x + y = 25$ at $(3, 4)$

```
RUN
EDIT IN YOUR FUNCTION F(X), THEN RUN 250

280        DEF FNF(X) =  (SQR(25 - (3 + X)^2) - 4)/X
RUN 250
EDIT IN YOUR POINT A, THEN RUN 270

290        A = 0
RUN 270
Random number seed (-32768 to 32767)? 5

WANT LEFTHAND OR RIGHTHAND LIMIT (L/R)? R
x                       f(x)

+0.2736567             -0.8064543
+0.0710505             -0.7640662
+0.0033764             -0.7506599
+0.0006197             -0.7501211
+0.0000224             -0.7500044
+0.0000009             -0.7500002
+0.0000004             -0.7500001
```

Figure 2.15   Run of Program LIMFCTN with the difference quotient in (7)

Figure 2.15 shows a run (in double-precision BASIC) of Program LIMFCTN with this difference quotient, and we see a modest rate of convergence to $-3/4 = -0.75$.

There is a way to speed up this convergence significantly. We must get the same value of the limit in (4), with $h \to 0$ through both positive and negative values. If we replace h with $-h$, we get

$$f'(x_0) = \lim_{h \to 0} \frac{f(x_0) - f(x_0 - h)}{h}. \tag{8}$$

If we now add equations (4) and (8) and divide by 2, the result is

$$f'(x_0) = \lim_{h \to 0} \frac{f(x_0 + h) - f(x_0 - h)}{2h}. \tag{9}$$

Thus

$$f'(x_0) = \lim_{h \to 0} D_s(h), \tag{10}$$

where

$$D_s(h) = \frac{f(x_0 + h) - f(x_0 - h)}{2h} \tag{11}$$

is the *symmetric* difference quotient of f at x_0.

In Example 1 the symmetric difference quotient is

$$D_s(h) = \frac{\sqrt{25 - (3 + h)^2} - \sqrt{25 - (3 - h)^2}}{2h}. \tag{12}$$

Figure 2.16 shows a double-precision run of Program LIMFCTN, and we see much more rapid convergence to the desired limit.

```
RUN
EDIT IN YOUR FUNCTION F(X), THEN RUN 250

280        DEF FNF(X) =
           (SQR(25 - (3 + X)^2) - SQR(25 - (3 - X)^2))/(2*X)
RUN 250
EDIT IN YOUR POINT A, THEN RUN 270

290        A = 0
RUN 270
Random number seed (-32768 to 32767)? 5

WANT LEFTHAND OR RIGHTHAND LIMIT (L/R)? R
x                      f(x)

+0.2736567             -0.7527650
+0.0710505             -0.7501850
+0.0033764             -0.7500004
+0.0006197             -0.7500000
+0.0000224             -0.7500000
+0.0000009             -0.7500000
+0.0000004             -0.7500000
```

Figure 2.16   Run of Program LIMFCTN with the symmetric difference quotient in (12)

To write a program for the specific purpose of computing derivatives, we take $x = a$ as the fixed point and write

$$f'(a) = \lim_{h \to 0} \frac{f(a + h) - f(a - h)}{2h} \tag{13}$$

using the symmetric difference quotient. Listing 2.17 shows Program DERIV. The function $f(x)$ and the fixed point a must be edited into lines 170 and 190, respectively. The value of the derivative is printed if and when the old value DOLD and the new value DNEW of the symmetric difference quotient differ by less than the error tolerance 0.000001.

```
100 REM--Program DERIV
110 REM--Computes the value of the derivative of the
115 REM--function f(x) defined in line 170, at the
120 REM--point x=a specified in line 190.
130 REM
140 REM--Initialization:
150 REM
160     DEFDBL A, D, F, H, X
170     DEF FNF(X) = 2^X
180     DEF FND(H) = (FNF(A + H) - FNF(A - H))/(2*H)
190     A = 0   :   H = 1   :   N = 0
200     DOLD = FND(1)
210 REM
220 REM--Compute derivative value:
230 REM
240     H = H/10   :   N = N + 1
250     DNEW = FND(H)
260     IF ABS(DNEW - DOLD) < .000001 THEN GOTO 310
270     IF N > 10 THEN
                PRINT "NO CONVERGENCE YET"   :   STOP
280     DOLD = DNEW
290     GOTO 240
300 REM
310     PRINT USING "DERIVATIVE = +#.#####"; DNEW
320     END
```

Listing 2.17   Program DERIV

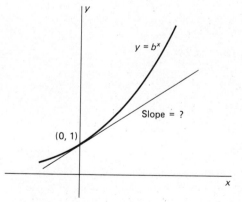

Figure 2.18   Tangent line to curve $y = b^x$ at $(1, 0)$

Example 2

Figure 2.18 shows the graph of the function $f(x) = b^x$ for $b > 0$. Find the slope $f'(0)$ of its tangent line at the point $(0, 1)$, for $b = 2$ and for $b = 3$.

Solution.   When we run Program DERIV with $f(x) = 2^x$ as listed, we get the
result DERIVATIVE $= +0.69315$. When we change line 170 to $f(x) = 3^x$,
we get the result DERIVATIVE $= +1.09861$. (The alert reader who knows how
to differentiate exponentials may notice that these two values are the natural
logarithms of 2 and 3, respectively.)

Since $f'(0) < 1$ for $b = 2$ and $f'(0) > 1$ for $b = 3$, it is natural to suspect that $f'(0)$ is precisely 1 for some value of b between 2 and 3. To see what
the significance of such a value of b would be, we differentiate the function
$f(x) = b^x$:

$$f'(x) = \lim_{h \to 0} \frac{b^{x+h} - b^x}{h}$$

$$= b^x \lim_{h \to 0} \frac{b^h - 1}{h} = f'(0)f(x). \tag{14}$$

It follows that if $f'(0) = 1$, then $f'(x) = f(x)$. That is, for such a value of b, *the
function $f(x) = b^x$ is its own derivative!*

The Number e

The number between 2 and 3 such that

$$f'(0) = \lim_{h \to 0} \frac{b^h - 1}{h} = 1 \tag{15}$$

is the famous number e that serves as the base number for natural logarithms. Listing 2.19 shows Program EFIND, which was written to compute
the value of e by solving equation (15) in the interval $[2, 3]$ by the method
of bisection that was discussed in Section 1.3. Note that Equation (15)
is of the form

$$G(b) = D(b) - 1 = 0, \tag{16}$$

where

$$D(b) = \lim_{h \to 0} \frac{b^h - 1}{h} = \lim_{h \to 0} \frac{b^h - b^{-h}}{2h}. \tag{17}$$

Since $G(2) < 0$ and $G(3) > 0$, we start with the interval $[2, 3]$ and repeatedly
bisect it, retaining at each step the subinterval on which $G(b)$ changes sign.

Lines 240–330 incorporate the basic bisection loop of Program BISECT
in Section 1.3. We could not simply run BISECT because we needed (in
essence) a multiline function definition to compute the value D of $f'(0)$ as
a function of b. This is done in the subroutine (starting at line 350) which
corresponds to Program DERIV and uses the symmetric difference
quotient defined in line 190. Figure 2.20 shows a run of Program EFIND

```
100 REM--Program EFIND
110 REM--Applies the method of repeated bisection to find
120 REM--value e = c between c1 = 2 and c2 = 3   such that
130 REM--the derivative of f(x) = c^x   at a = 0   is 1.
140 REM
150 REM--Initialization:
160 REM
170     DEFDBL A, B, C, D, F, H, M, X
180     DEF FNF(X) = B^X
190     DEF FND(B) = (FNF(A+H) - FNF(A-H))/(2*H)
200     C1 = 2   :    C2 = 3
210 REM
220 REM--Repeated bisection loop:
230 REM
240     PRINT USING "#.######        "; C1, C2
250     B = C1   :   GOSUB 350
260     DC1 = D - 1
270     M = (C1 + C2)/2
280     B = M    :   GOSUB 350
290     DM = D - 1
300     IF DC1*DM < 0 THEN LET C2 = M ELSE C1 = M
310     IF C2 - C1 > .000001 THEN GOTO 240
320     PRINT : PRINT USING " e = #.#####"; (C1 + C2)/2
330     STOP
340 REM
350 REM--Subroutine DERIV:
360 REM
370     A = 0    :    H = 1
380     DOLD = FND(B)
390     H = H/10
400     DNEW = FND(B)
410     IF ABS(DNEW - DOLD) < .0000001 THEN GOTO 440
420     DOLD = DNEW
430     GOTO 390
440     D = DNEW
450     RETURN
460 REM
470     END
```

Listing 2.19   Program EFIND

```
RUN
2.000000      3.000000
2.500000      3.000000
2.500000      2.750000
2.625000      2.750000
2.687500      2.750000
2.687500      2.718750
2.703125      2.718750
2.710938      2.718750
2.714844      2.718750
2.716797      2.718750
2.717773      2.718750
2.718262      2.718750
2.718262      2.718506
2.718262      2.718384
2.718262      2.718323
2.718262      2.718292
2.718277      2.718292
2.718277      2.718285
2.718281      2.718285
2.718281      2.718283

e = 2.71828
```

Figure 2.20   Run of Program EFIND

(in double-precision BASIC) that yields the value $e = 2.71828$ that is accurate to five decimal places.

PROBLEMS

In each of Problems 1–5, use Program DERIV to compute the derivative $f'(a)$ of the given function $f(x)$ at the indicated point a. In each case the actual value is a rational number p/q, where the integers p and q are both numerically less than 10.

1. $f(x) = \dfrac{1}{x}; \ a = 2$

2. $f(x) = \sqrt{2x + 1}; \ a = 4$

3. $f(x) = (x + 3)^{4/3}; \ a = \dfrac{3}{8}$

4. $f(x) = \ln (7x + 1); \ a = 1$

5. $f(x) = \tan^{-1} x \ [= \text{ATN(X) in BASIC}]; \ a = -\dfrac{1}{2}$

6. It can be shown that the *second* derivative $f''(a)$ of the function $f(x)$ at the point $x = a$ is given by

$$f''(a) = \lim_{h \to 0} \frac{f(a + h) - 2f(a) + f(a - h)}{h^2}.$$

Alter Program DERIV so as to obtain a program that computes $f''(a)$. Check it with the function $f(x) = 1/x$, for which $f''(1) = 2$.

7. It is known that

$$e^x = \lim_{n \to \infty} \left(1 + \frac{x}{n}\right)^n$$

for all x. It follows that the natural logarithm $\ln a$ of the number $a > 0$ is the solution x of the equation

$$\lim_{n \to \infty} \left(1 + \frac{x}{n}\right)^n = a.$$

Write a program that when a is input will solve this equation by the method of bisection to compute $\ln a$. This program will be much like Program EFIND, except that your subroutine will compute the limit as $n \to \infty$ of $(1 + x/n)^n$. Check to see that your program yields $\ln 2 \approx 0.69315$.

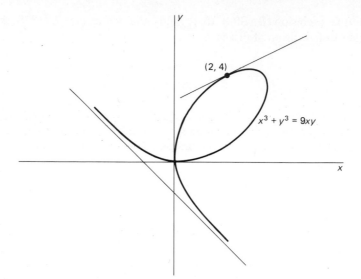

Figure 2.21   The folium of Descartes

8. Figure 2.21 shows the "folium of Descartes," whose equation is $x^3 + y^3 = 9xy$. Write a program to compute the slope of the tangent line to this curve at the point $(2, 4)$. This slope will be

$$m = \lim_{h \to \infty} \frac{\Delta y}{\Delta x} = \lim_{h \to 0} \frac{y_1 - 4}{h},$$

where y_1 denotes the largest of the three roots of the cubic equation

$$(2 + h)^3 + y^3 = 9(2 + h)y$$

in y that is obtained by substituting $x = 2 + h$. You can solve this cubic equation in a subroutine that uses Newton's method, starting with the initial estimate $y = 4$.

9. Write a program to find the point on the folium of Descartes where the tangent line is horizontal.

2.4 NEWTON'S METHOD REVISITED

In Section 1.5 we discussed Newton's method of constructing a rapidly convergent sequence of successive approximations to a solution r of a *polynomial* equation $f(x) = 0$. Now that derivatives of more general functions are available, we can apply Newton's method to *transcendental* equations—ones that involve exponential, logarithmic, or trigonometric functions.

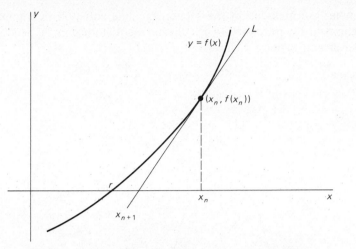

Figure 2.22   Geometric interpretation of Newton's method

Figure 2.22 illustrates a geometric derivation of Newton's iterative formula. Given the nth approximation x_n to the solution r of the equation

$$f(x) = 0, \tag{1}$$

the tangent line to the curve $y = f(x)$ at the point $(x_n, f(x_n))$ is used as follows to construct the next approximation, x_{n+1}. We begin at the point x_n on the x-axis, and go vertically up (or down) to the point $(x_n, f(x_n))$ on the curve. Then we follow the tangent line L to the point where it intersects the x-axis. That point will be x_{n+1}.

We obtain Newton's iterative formula for x_{n+1} by computing the slope of the tangent line L in two different ways: from the derivative and from the two-point definition of slope. The first gives $f'(x_n)$, and inspection of Figure 2.22 yields $[f(x_n) - 0]/(x_n - x_{n+1})$. Equating these two values, we get the equation

$$f'(x_n) = \frac{f(x_n) - 0}{x_n - x_{n+1}},$$

which is readily solved for the iterative formula

$$x_{n+1} = x_n - \frac{f(x_n)}{f'(x_n)} \tag{2}$$

of Newton's method. To apply this formula as we did in Section 1.5, we need only know how to find the derivative $f'(x)$ of the function $f(x)$ that appears in Equation (1). The derivatives that are needed in the following two

examples are included in the section on differentiation of trigonometric
functions in any introductory calculus textbook (for example, Section 3.6 of
Edwards and Penney, *Calculus*).

Example 1

Find all solutions of the equation

$$x + 3 \cos x = 1. \tag{3}$$

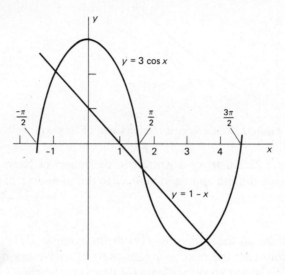

Figure 2.23

Solution.   To investigate the solutions of Equation (3), we rewrite it in the form

$$3 \cos x = 1 - x.$$

Figure 2.23 shows the graphs $y = 3 \cos x$ and $y = 1 - x$, whose three points
of intersection tell us that our equation has three real solutions. By inspection of
the figure, we see that they lie roughly at $x = -0.75$, 1.75, and 3.50.

To apply Newton's method we write Equation (3) in the form

$$f(x) = x + 3 \cos x - 1 = 0 \tag{4}$$

and note that

$$f'(x) = 1 - 3 \sin x. \tag{5}$$

Listing 2.24 shows Program NEWTON as it appeared initially in Listing 1.24
in Section 1.5, except that now the function $f(x)$ in (4) and its derivative in (5)
have been edited into lines 230 and 240 of the program. Figure 2.25 shows
the results of three runs of Program NEWTON with the three initial values

```
100 REM--Program NEWTON
110 REM--Applies Newton's method to solve the equation
120 REM--f(x) = 0.  Lines 230 and 240 must be edited to
130 REM--read FCTN = f(X) and DERIV = f'(X) respectively.
140 REM--The initial guess is input in line 160.
150 REM
160       INPUT "INITIAL GUESS X1"; X
170       N = 1
180       PRINT "N", "  X" : PRINT
190 REM
200 REM--Newton's iteration:
210 REM
220       PRINT N, X
230       FCTN  = X + 3*COS(X) - 1
240       DERIV = 1 - 3*SIN(X)
250       IF DERIV = 0 THEN PRINT
              "ZERO DERIVATIVE--TRY NEW INITIAL GUESS" :
              GOTO 160
260       XNEW = X - FCTN/DERIV
270       IF ABS(XNEW - X) < .000001*ABS(X) THEN
              PRINT "SOLUTION = ";XNEW : STOP
280       X = XNEW : N = N + 1
290       GOTO 220
300 REM
310       END
```

Listing 2.24   Program NEWTON

```
RUN
INITIAL GUESS X1? -.75
N              X

1                 -.75
2                 -.8961671
3                 -.8894831
4                 -.8894704
SOLUTION = -.8894704
```

```
RUN
INITIAL GUESS X1?  1.75
N              X

1                 1.75
2                 1.86028
3                 1.862364
SOLUTION =  1.862365
```

```
RUN
INITIAL GUESS X1?  3.5
N              X

1                 3.5
2                 3.65074
3                 3.638045
4                 3.637958
SOLUTION =  3.637958
```

Figure 2.25 The three solutions of Equation (3)

indicated above. Thus Equation (3) has the three solutions $x \approx -0.88947$, 1.86237, and 3.63796.

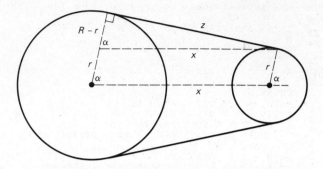

Figure 2.26   The belt and pulleys of Example 2

Example 2

Figure 2.26 shows a belt of total length L stretched tightly around two circular pulleys with radii R and r, with x denoting the distance between their centers. Using the indicated angle α, we see that

$$L = R(2\pi - 2\alpha) + 2z + 2r\alpha, \tag{6}$$

because an angle θ (radians) in a circle of radius a subtends an arc of length $s = a\theta$. From the right triangle in the figure we see that

$$x = (R - r)\sec\alpha, \tag{7}$$

$$z = (R - r)\tan\alpha, \tag{8}$$

and substitution of (8) into (6) yields

$$L = 2(R - r)(\tan\alpha - \alpha) + 2\pi R. \tag{9}$$

Now suppose that—given R, r, and L—we want to find x. We first solve Equation (9) using Newton's method, and find x from (7). For instance, if $R = 3$ ft, $r = 1$ ft, and $L = 25$ ft, then Equation (9) becomes

$$f(\alpha) = 4\tan\alpha - 4\alpha + 6\pi - 25 = 0, \tag{10}$$

and then

$$f'(\alpha) = 4\sec^2\alpha - 4 = \frac{4}{\cos^2\alpha} - 4. \tag{11}$$

Figure 2.27 shows the result of editing lines 230 and 240 of Program NEWTON accordingly (with x denoting α in the program), and then running it with initial guess X1 = 1. We find the solution $\alpha = 1.223295$, so the distance between the centers of the two pulleys is [using (7)]

$$x = 2 \sec (1.223295) \approx 5.87 \text{ ft.}$$

```
230        FCTN  = 4*TAN(X)  - 4*X + 6*3.141593 - 25
240        DERIV = 4/(COS(X)^2) - 4

INITIAL GUESS X1?   1
N                    X

1                    1
2                    1.404121
3                    1.319146
4                    1.250879
5                    1.225641
6                    1.223312
7                    1.223295
SOLUTION =    1.223295
```

Figure 2.27   Solving Equation (10)

PROBLEMS

In each of Problems 1-7, use Program NEWTON to find a solution of the given equation in the interval indicated.

1. $x = \cos x$; $[0, 2]$

2. $2x = \cos x$; $[0, 1]$

3. $x^2 = \sin x$; $\left[\dfrac{1}{2}, 1\right]$

4. $4x = 4 + \sin x$; $[1, 2]$

5. $5x + \cos x = 5$; $[0, 1]$

6. $x + \tan x = 0$; $[2, 3]$

7. $x + \tan x = 0$; $[11, 12]$

In each of Problems 8-10, use the method of Example 1 to find all real roots of the given equation.

8. $x^2 = \cos x$

9. $x = 2 \sin x$

10. $x + 5 \cos x = 0$

11. A moon of a certain planet has an elliptical orbit with eccentricity $1/2$, and its period of revolution about the planet is 100 days. If the moon is at the position $(a, 0)$ when $t = 0$, then, as illustrated in Figure 2.28, its central angle θ after t days is given by Kepler's equation,

$$\frac{2\pi t}{100} = \theta - \frac{1}{2} \sin \theta.$$

Use Newton's method to solve for θ when $t = 17$ days.

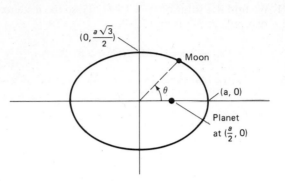

Figure 2.28   The elliptical orbit of Problem 11

12. The equation $x + \tan x = 0$ is important in a variety of applications—for example, in the study of the diffusion of heat. It has a sequence $x_1, x_2, x_3, \ldots$ of positive roots, with the nth one slightly larger than $(n - 1/2)\pi$. Use Newton's method to compute the first four solutions, $x_1, x_2, x_3,$ and x_4.

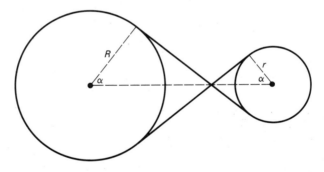

Figure 2.29   The belt and pulleys of Problem 13

13. Figure 2.29 shows a belt of length L around two pulleys with radii R and r as in Example 2, except that the belt crosses itself once. Find the distance x between the centers of the two pulleys if $L = 30$ ft, $R = 3$ ft, and $r = 1$ ft. *Suggestion:* Show first that

$$x = (R + r) \sec \alpha,$$
$$L = 2(R + r)(\pi - \alpha + \tan \alpha).$$

14. Write a version of Program NEWTON that, instead of defining the derivative of $f(x)$ as in line 240, uses a subroutine corresponding to Program DERIV (Listing 2.17) to compute $f'(x)$.

2.5 MAXIMUM-MINIMUM PROBLEMS

One of the staple features of introductory calculus is the solution of maximum-minimum problems. In this section we discuss computer solutions
of such problems. The following example from Section 3.5 of Edwards and
Penney, *Calculus* is typical.

Example 1

A piece of sheet metal is rectangular, 5 ft wide and 8 ft long. Equal squares are to
be cut from its corners and the resulting piece of metal folded and welded to form
a box with an open top, as shown in Figure 2.30. How should this be done to get a
box of the largest possible volume?

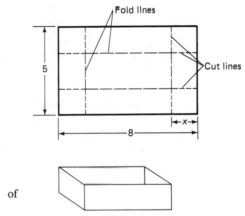

Figure 2.30   Making the box of
Example 1

To analyze this problem, we denote by x the edge length of each corner
square that is removed. The shape and the volume V of the box constructed
are determined by x. Since the width of the sheet is 5 ft, we can choose any
value of x in the interval $0 \leqq x \leqq 2.5$, although the endpoints $x = 0$ and
$x = 2.5$ correspond to "degenerate" boxes that have zero volume. Consulting
Figure 2.30, we see that the box will have height x, and its base will measure
$5 - 2x$ by $8 - 2x$. Hence its volume is given by

$$V = x(5 - 2x)(8 - 2x),$$
$$V(x) = 4x^3 - 26x^2 + 40x \qquad (1)$$

as a function of x. Consequently, Example 1 amounts to the purely mathematical problem of determining what value of x in the closed interval $[0, 2.5]$
produces the maximum value of the function $V(x)$ defined in (1).

More generally, consider the problem of finding the maximum and
minimum values of the given function $f(x)$ on the interval $[a, b]$. Pictures
like Figure 2.31 suggest that each extreme value will always occur *either* at

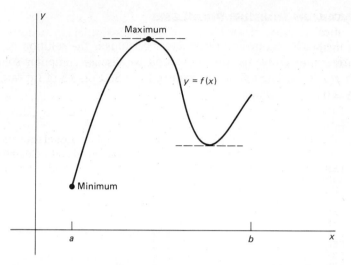

Figure 2.31   Maximum and minimum values of $f(x)$ on $[a, b]$

one of the two endpoints *or* at an interior point of the interval where the tangent line is horizontal, and therefore has slope $f'(x) = 0$. Indeed, the following theorem is proved in Section 3.4 of Edwards and Penney, *Calculus*.

     Maximum-Minimum Theorem.   Suppose that the function $f(x)$ is differentiable on the closed interval $[a, b]$. Then $f(x)$ attains its maximum value (and likewise its minimum value) on $[a, b]$ *either* at one of the endpoints a and b *or* at an interior point of the interval where $f'(x) = 0$.

     For instance, let $f(x) = 4x^3 - 26x^2 + 40x$ be the volume function in Equation (1). The theorem tells us that its maximum value on $[0, 2.5]$ occurs either at $x = 0$ or at $x = 2.5$ (the two endpoints) or at an interior point where the derivative

$$f'(x) = 12x^2 - 52x + 40$$
$$= 4(3x - 10)(x - 1) = 0. \tag{2}$$

The only points where $f'(x) = 0$ are $x = 1$ and $x = 10/3$, and the latter does not lie in the interval $[0, 2.5]$. Thus the only candidates for a maximum value are at $x = 0$, $x = 1$, and $x = 2.5$. Evaluating the function $f(x)$ at each of these points, we find that

$$f(0) = 0, \quad f(1) = 18, \quad f(2.5) = 0.$$

Hence the box with maximal volume $V = 18$ ft^3 is constructed with $x = 1$ ft. Its base dimensions are $5 - 2x = 3$ ft by $8 - 2x = 6$ ft.

Once an applied problem as in Example 1 has been translated into the mathematical question of what is the maximum or minimum value of a particular function $f(x)$ on a specified interval $[a, b]$, the remaining problem is to solve the equation $f'(x) = 0$. If this can be done, we then evaluate the function $f(x)$ at the endpoints $x = a$ and $x = b$ and at those solution points of $f'(x) = 0$ that lie between a and b. By visual inspection of the resulting values of the function we can finally determine the desired maximum or minimum value.

```
100 REM--Program MAXMIN
110 REM--Searches for the possible maximum and minimum
115 REM--values of the function f(x) in the input
120 REM--interval [x1,x2].  The program pauses during
130 REM--execution for the function, its derivative,
140 REM--and its second derivative to be edited into
150 REM--lines 260, 270, and 280 respectively.
160 REM
170 REM--Initialization:
180 REM
190         PRINT "EDIT IN YOUR FUNCTION F(X), THEN RUN 210"
200         PRINT   :   EDIT 260
210         PRINT "EDIT IN ITS DERIVATIVE D(X), THEN RUN 230"
220         PRINT   :   EDIT 270
230         PRINT
            "EDIT IN ITS 2ND DERIVATIVE DD(X), THEN RUN 260"
240         PRINT
250         EDIT 280
260         DEF  FNF(X) =
270         DEF  FND(X) =
280         DEF FNDD(X) =
290         PRINT  :  INPUT "ENDPOINTS X1, X2"; X1, X2   :
            PRINT
300         DIM A(100)
310         A(0) = X1  :  J = 1  :  X = X1  :
                          H = (X2 - X1)/100
320         PRINT "COMPUTATION IN PROGRESS"  : PRINT
330 REM
340 REM--Loop to find critical points:
350 REM
360         FOR I = 1 TO 100
370             IF FND(X)*FND(X+H) < 0
                    THEN XNEW = X  :  GOSUB 510
380             X = X + H
390             IF FND(X) = 0
                    THEN A(J) = X  :  J = J + 1
400         NEXT I
410         A(J) = X2
420 REM
430 REM--Print possible maxima and minima:
440 REM
450         PRINT "x", "f(x)"  :  PRINT
460         FOR I = 0 TO J
470             PRINT A(I), FNF(A(I))
480         NEXT I
490         STOP
500 REM
510 REM--Newton's method subroutine:
520 REM
```

Listing 2.32   Program MAXMIN

```
530        XOLD = XNEW
540        XNEW = XOLD - FND(XOLD)/FNDD(XOLD)
550        IF ABS(XNEW - XOLD) > .000001*ABS(XNEW)
           THEN GOTO 530
560        A(J) = XNEW    :    J = J + 1
570        RETURN
580 REM
590        END
```

Listing 2.32 (con't.)

Program MAXMIN shown in Listing 2.32 was written to implement the procedure described above. During execution the program prompts us to edit into lines 260, 270, and 280 the function $f(x) = F(X)$, its derivative $f'(x) = D(X)$, its second derivative $f''(x) = DD(X)$, and to input the interval $[x_1, x_2]$ on which we wish to find the maximum and minimum values of $f(x)$. The reason we need the *second* derivative is that we are going to use Newton's method to solve the equation $f'(x) = 0$ for the possible maximum-minimum points.

In the loop consisting of lines 360–400, the interval $[x_1, x_2]$ is first subdivided into 100 equal subintervals of the form $[x, x + h]$. Line 370 checks to see whether $f'(x)$ changes sign in $[x, x + h]$. If so, this interval contains a solution of the equation $f'(x) = 0$, and we use the Newton's method subroutine in lines 530–570 to find it with relative accuracy less than 0.000001. The solutions found in this way are stored in the array A(J). Line 390 checks whether any of the points of subdivision are coincidentally solutions of $f'(x) = 0$. All of these points—together with the two endpoints x_1 and x_2, which are placed in the same array by lines 310 and 410—constitute the possibilities for a maximum or minimum value of the function $f(x)$ on the interval $[x_1, x_2]$.

Finally, lines 450–480 print out the values x and $f(x)$ for all the possibilities that we have found. It should be noted that this program is not completely "fail-safe." In the unlikely event that a single one of our 100 subintervals happened to contain two distinct solutions of $f'(x) = 0$, we would find only one of them, and might thereby miss the extreme value we are looking for.

Example 2

Find the maximum and minimum values attained by the function

$$f(x) = 3x^4 - 2x^3 - 18x^2 \tag{3}$$

on the closed interval $[-2, 3]$.

Solution. Figure 2.33 shows the results of a run of Program MAXMIN with the function $f(x)$, its derivative $f'(x) = 12x^3 - 6x^2 - 36x$, and its second derivative $f''(x) = 36x^2 - 12x - 36$ edited in. Scanning the column of

```
RUN
EDIT IN YOUR FUNCTION F(X), THEN RUN 210

260       DEF FNF(X) = 3*X^4 - 2*X^3 - 18*X*X
RUN 210
EDIT IN ITS DERIVATIVE D(X), THEN RUN 230

270       DEF FND(X) = 12*X^3 - 6*X*X - 36*X
RUN 230
EDIT IN ITS 2ND DERIVATIVE DD(X), THEN RUN 260

280       DEF FNDD(X) = 36*X*X - 12*X - 36
RUN 260

ENDPOINTS X1, X2? -2, 3

COMPUTATION IN PROGRESS

X                 f(x)

-2                -8
-1.5              -18.5625
0                 0
2                 -40
3                 27
```

Figure 2.33   Run of Program MAXMIN

values of $f(x)$, we see that the maximum value $f(3) = 27$ occurs at an endpoint, while the minimum value $f(2) = -40$ occurs at an interior point of $[-2, 3]$.

Use of Graphics

In Section 1.6 we discussed the use of computer graphics to locate roots of equations by visual inspection. Maxima and minima of functions can be located similarly. Program GRAPH2, shown in Listing 2.34, is a slight alteration of our earlier graphing program (GRAPH, Listing 1.37 in Section 1.6). It plots the graph on the input interval [XMIN, XMAX] of the function $f(x)$ that is edited into line 160. By rerunning Program GRAPH2 with successively smaller input intervals, we effectively can magnify the graph $y = f(x)$ in a neighborhood of an extremum so as to "eyeball" the maximum or minimum value.

```
100 REM--Program GRAPH2
110 REM--Plots graph of the function y = f(x)
115 REM--on the input interval [XMIN,XMAX].  The
120 REM--function f(x) must be defined in line 160.
130 REM
140       PRINT "EDIT IN YOUR FUNCTION F(X), THEN RUN 160"
150       PRINT  :  EDIT 160
160       DEF FNF(X) =
170       INPUT "XMIN, XMAX"; XMIN, XMAX
180 REM
190 REM--Set scale on vertical axis:
```

Listing 2.34   Program GRAPH2

```
200 REM
210       YMAX = FNF(XMIN) : YMIN = FNF(XMIN)
220       FOR I=1 TO 20
230           X = XMIN + I*(XMAX - XMIN)/20
240           IF YMAX < FNF(X) THEN YMAX = FNF(X)
250           IF YMIN > FNF(X) THEN YMIN = FNF(X)
260       NEXT I
270       DX = XMAX - XMIN : DY = YMAX - YMIN
275 REM
280 REM--Define viewing transformation:
290 REM
300       UMIN = 150 : UMAX = 550  :  DU = 400
310       VMIN = 19  : VMAX = 179  :  DV = 160
320       DEF FNU(X) = UMIN + (X - XMIN)*DU/DX
330       DEF FNV(Y) = VMAX - (Y - YMIN)*DV/DY
340 REM
350 REM--Draw and label axes:
360 REM
370       SCREEN 2 : CLS : KEY OFF
380       LINE (100,  2) - (100,190)
390       LINE ( 95, 19) - (105, 19)
400       LINE ( 95,179) - (105,179)
410       LINE ( 75, 99) - (625, 99)
420       FOR I=0 TO 10
430           LINE (150 + 40*I,  94)
                - (150 + 40*I, 104)
440       NEXT I
450       K  = 14
460       LOCATE  1,14 : PRINT "y"
470       LOCATE K,78 : PRINT "x"
480       LOCATE  3,4  : PRINT USING "+##.####"; YMAX
490       LOCATE 23,4  : PRINT USING "+##.####"; YMIN
500       LOCATE K,18 : PRINT XMIN
510       LOCATE K,43 : PRINT (XMIN + XMAX)/2
520       LOCATE K,68 : PRINT XMAX
530 REM
540 REM--Plot graph of function:
550 REM
560       X = XMIN : H = (XMAX - XMIN)/200
570       FOR I=0 TO 200
580           Y = FNF(X)
590           PSET (FNU(X), FNV(Y))
600           X = X + H
610       NEXT I
620       LOCATE 1,1
630 REM
640       END
```

Listing 2.34 (con't)

Example 3

Find the minimum value attained by the function

$$f(x) = 4x^4 - 11x^2 + 3x - 6. \qquad (4)$$

Solution. We seek a minimum value (only) because it is clear that $f(x) \longrightarrow \infty$ as $x \longrightarrow \pm\infty$. Indeed, it is apparent that the leading term $4x^4$ in (4) predominates when $|x| > 2$, because $4(2)^4 > 11(2)^2 + 3(2) + 6$. We therefore first run GRAPH2 to plot $y = f(x)$ on the interval $[-2, 2]$. From Figure 2.35 it is obvious that the minimum value is attained somewhere in the interval $[-1.5, -1.0]$. Figures 2.36 through 2.38 show the results of subsequent runs using the successively smaller intervals $[-1.5, -1.0]$, $[-1.3, -1.2]$, and $[-1.24, -1.23]$. From Figure 2.38 it appears that a minimum value of about -17.1772 is attained somewhere between -1.236 and -1.235. This is as much accuracy as graphs can provide with the resolution available. However, now that we know the approximate location of the extremum, we can use Program MAXMIN for greater accuracy. The run shown in Figure 2.39 shows that the minimum value is

$$f(-1.235616) \approx -17.17724.$$

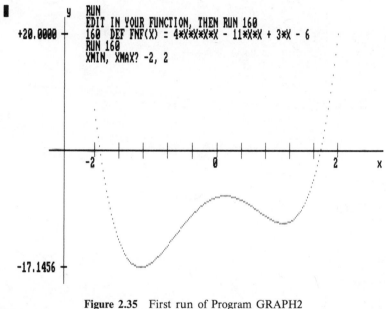

Figure 2.35   First run of Program GRAPH2

PROBLEMS

In each of Problems 1–8, use Program MAXMIN to find the maximum and minimum values of the given function $f(x)$ on the interval indicated.

1. $f(x) = 5 - 12x - x^2$ on $[-1, 1]$
2. $f(x) = x^3 - 3x^2 - 9x + 5$ on $[-2, 4]$
3. $f(x) = 3x^5 - 5x^3$ on $[-2, 2]$
4. $f(x) = 50x^3 - 105x^2 + 72x$ on $[0, 1]$
5. $f(x) = x^{10} - 5x^2 + 5$ on $[0, 2]$

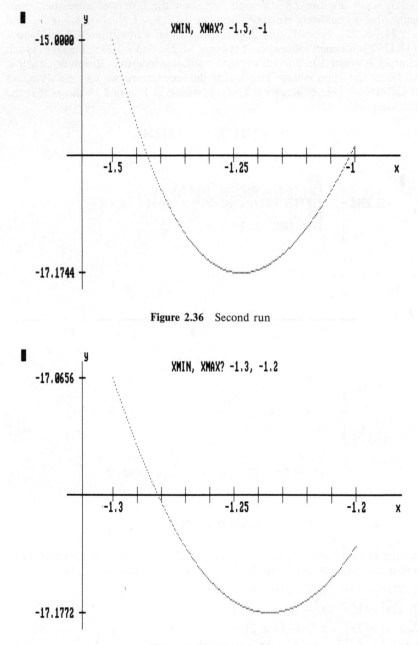

Figure 2.36 Second run

Figure 2.37 Third run

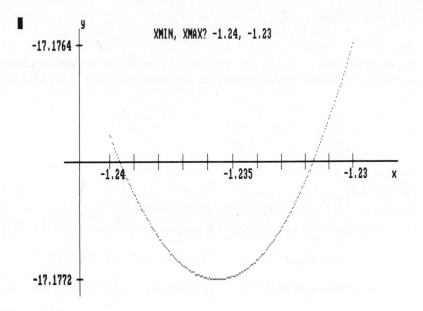

Figure 2.38   Fourth run

```
RUN
EDIT IN YOUR FUNCTION F(X), THEN RUN 210

260      DEF FNF(X) = 4*X*X*X*X - 11*X*X + 3*X - 6
RUN 210
EDIT IN ITS DERIVATIVE D(X), THEN RUN 230

270      DEF FND(X) = 16*X*X*X - 22*X + 3
RUN 230
EDIT IN ITS 2ND DERIVATIVE DD(X), THEN RUN 260

280      DEF FNDD(X) = 48*X*X - 22
RUN 260

ENDPOINTS X1, X2? -1.24, -1.23

COMPUTATION IN PROGRESS

x                f(x)

-1.24            -17.17675
-1.235616        -17.17724
-1.23            -17.17644
```

Figure 2.39   Run of Program MAXMIN for Example 3

6. $f(x) = |x - 3|$ on $[0, 5]$

7. $f(x) = \dfrac{\ln x}{x}$ on $[1, 10]$

8. $f(x) = xe^{-x}$ on $[0, 5]$

In each of Problems 9–14, use the method of Example 3 to find the (absolute) maximum or minimum value–whichever is appropriate–of the given function $f(x)$.

9. $f(x) = 3 - |x - 2|$

10. $f(x) = 3x^4 - 4x^3 - 12x^2$

11. $f(x) = 6 + 8x^2 - x^4$

12. $f(x) = 10 + 32x - x^4$

13. $f(x) = x^{4/3} - 4x^{1/3}$

14. $f(x) = x^{2/3}(x^2 - 2x - 6)$

15. The volume V of 1 kg of water at a temperature T between 0 and 30°C is

$$V = 999.87 - (0.06426)T + (0.0085043)T^2 - (0.0000679)T^3$$

cubic centimeters. At what temperature does water have its maximal density?

16. A cylindrical can with a volume of 125 in.³ is made by cutting the top and bottom out of metal squares and forming the curved (lateral) surface by bending a rectangular sheet of metal to meet at its ends. If r denotes the radius of the can, the total area A of the material used (the two squares and the rectangle) is given by

$$A = 8r^2 + \frac{250}{r}.$$

What value of r minimizes A?

17. Suppose we know from a graph that x_0 lies just to the left of a local minimum of the function $f(x)$. Write a program to implement the following search procedure to find the minimum. Starting with a given "step size" $h > 0$, we calculate $f(x)$ at the points $x_1 = x_0 + h$, $x_2 = x_0 + 2h$, $x_3 = x_0 + 3h, \ldots$ successively as long as $f(x_n) < f(x_{n-1})$. But the first time it happens that $f(x_n) \geq f(x_{n-1})$, we replace x_0 with x_n and h by $-h/10$ and then start "stepping" back to the left. As soon as we have gone too far to the left, we start back to the right with step size $h/100$. After k iterations of this process we know the location of the minimum with accuracy $h/10^k$.

Area and the Integral

3

3.1 ELEMENTARY AREA COMPUTATIONS

In Section 2.3 we saw that the problem of defining the tangent line at a point of the graph $y = f(x)$ leads to the concept of the derivative $f'(x)$ of the function $f(x)$. The value $f'(x)$ equals the slope of the tangent line at the point $(x, f(x))$ on the curve. Although its definition is motivated geometrically, the importance of the derivative in practical applications lies more frequently in its interpretation as the *rate of change* of the functional value $f(x)$ with respect to the independent variable x.

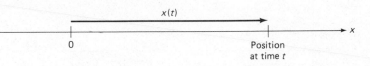

Figure 3.1   A particle moving along the x-axis

For instance, suppose that $x = x(t)$ denotes the position at time t of a particle that is moving along the x-axis (Figure 3.1). In this situation, time t is the independent variable and position x is the dependent variable. The rate of change of the particle's position x with respect to time t is called its *velocity*,

$$v = x'(t). \tag{1}$$

Thus the particle's velocity $v(t)$ at time t is the slope of the tangent line at the point $(t, x(t))$ to the curve $x = x(t)$ in the tx-plane (see Figure 3.2). We find the velocity function $v(t)$ by differentiating the position function $x(t)$.

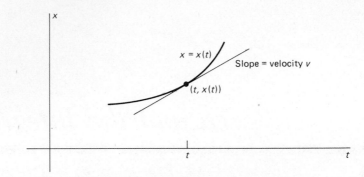

Figure 3.2   Velocity as rate of change of position

Now let's consider the reverse problem: If the velocity function $v(t)$ is known, how do we find the position function $x(t)$? For instance, we might be in a submarine traveling beneath the polar icecap. Its speedometer records the velocity function $v(t)$, and we want to calculate how far it has gone at time t.

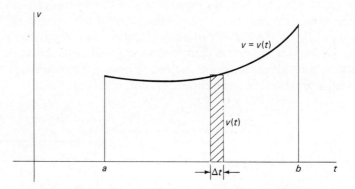

Figure 3.3   Distance traveled as area under the velocity curve

To answer this question, we consider the graph $v = v(t)$ in the tv-plane (Figure 3.3). Let $[t, t + \Delta t]$ be such a short time interval that the velocity $v(t)$ does not change appreciably between time t and time $t + \Delta t$. Then the distance traveled during this short time interval is given (to a sufficiently close approximation) by the product $v(t) \, \Delta t$, which in turn equals the area of the shaded rectangular strip in Figure 3.3. Now look at the region that lies under the curve $v = v(t)$ and above the interval

$a \leq t \leq b$. If we consider this region to consist of a large number of narrow vertical strips like the one shown, then we conclude that its total area should equal the total distance $x(b) - x(a)$ traveled by the particle between time $t = a$ and time $t = b$:

$$x(b) - x(a) = \text{area under curve } v = v(t) \text{ over the interval } [a, b]. \qquad (2)$$

Therefore, the problem of finding the position function $x(t)$ from the velocity function $v(t)$ reduces to the problem of computing the *area* of a region in the plane.

This relationship between position and velocity is a particular instance of the relationship between a function $F(x)$ and its derivative $f(x) = F'(x)$. Assuming that the derivative or rate of change $f(x)$ is positive valued, the increase $F(b) - F(a)$ in the value of $F(x)$, from $x = a$ to $x = b$, equals the area A of the region in the xy-plane that lies under the curve $y = f(x)$ and above the interval $[a, b]$. In calculus this area is denoted by

$$A = \int_a^b f(x) \, dx, \qquad (3)$$

and is called the *integral* of the function $f(x)$ from $x = a$ to $x = b$.

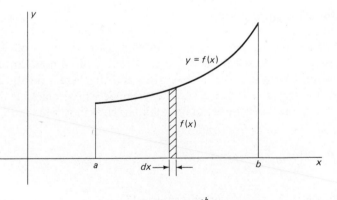

Figure 3.4   The area $A = \displaystyle\int_a^b f(x) \, dx$

Figure 3.4 illustrates the origin of the notation in (3). The strip shown is regarded as a rectangle with height $f(x)$, infinitesimal width dx, and hence area $f(x) \, dx$. The total area A is thought of as a sum of such strip areas—the integral sign $\int$ originally was an elongated gothic script S for *summa*. Although the integral can (and must!) be given a purely analytical definition in calculus, it will suffice here for us to regard it as simply a handy notation for area.

According to the *fundamental theorem of calculus* that is discussed and proved in Section 5.5 of Edwards and Penney, *Calculus*, the function $F(x)$ and its derivative $f(x) = F'(x)$ satisfy the relation

$$F(b) - F(a) = \int_a^b f(x)\, dx \tag{4}$$

Because $x'(t) = v(t)$, the formula

$$x(b) - x(a) = \int_a^b v(t)\, dt, \tag{5}$$

which expresses (2) in integral notation, is a special case of this theorem.

In short, the change $F(b) - F(a)$ in the function $F(x)$ from $x = a$ to $x = b$ equals the integral of its rate of change or derivative $f(x) = F'(x)$. If we regard the single value $F(a)$ as known—analogous to knowing the initial position of a moving particle—then Formula (4) tells how to find any other value $F(b)$ by calculating an appropriate area under the curve $y = f(x)$.

With this brief introduction to the integral, the remainder of this chapter is devoted to techniques for approximating the integral $\int_a^b f(x)\, dx$, that is, the area under the given curve $y = f(x)$ from $x = a$ to $x = b$.

The Midpoint Approximation

Perhaps the simplest way to approximate the integral of $f(x)$ from $x = a$ to $x = b$ is to subdivide the interval $[a, b]$ into n equal subintervals each having length $h = \Delta x = (b - a)/n$, and then erect on the ith subinterval (for each $i = 1, 2, 3, \ldots, n$) a rectangle whose height is the value $f(m_i)$ of f at the midpoint m_i of the subinterval, as illustrated in Figure 3.5. Thus

$$m_i = \frac{1}{2}(x_{i-1} + x_i) \qquad \text{where } x_i = a + ih \tag{6}$$

for each $i = 1, 2, \ldots, n$, and the area of the ith rectangle is

$$\Delta A_i = f(m_i)h. \tag{7}$$

Pictures like Figure 3.5 suggest that if n is sufficiently large (and hence h is small), then the sum

$$\sum_{i=1}^n \Delta A_i = f(m_1)h + f(m_2)h + \cdots + f(m_n)h \tag{8}$$

is a good approximation to the actual area $A = \int_a^b f(x)\, dx$ under the curve

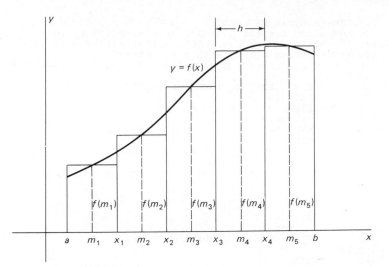

Figure 3.5 The rectangles comprising the midpoint approximation (with $n = 5$ subintervals)

$y = f(x)$ from $x = a$ to $x = b$. Indeed, it can be proved that if f is continuous on $[a, b]$, then

$$\int_a^b f(x)\, dx = \lim_{n \to \infty} \sum_{i=1}^{n} f(m_i)h. \tag{9}$$

This fact motivates the definition of the *midpoint approximation*

$$\int_a^b f(x)\, dx \approx h \sum_{i=1}^{n} f(m_i) \tag{10}$$

to the integral using n equal subintervals of length $h = (b - a)/n$.

Listing 3.6 shows Program MIDPOINT, which computes the midpoint approximation in (10). When this program is run, we begin by editing our function $f(x)$ into line 190. Then the endpoints a and b of our interval and the number n of subintervals we want are input at lines 200 and 220. The loop in lines 240–300 computes and prints the midpoint sum (10). Although double-precision arithmetic is specified in line 180, line 300 prints only five decimal places; this formatting instruction can be changed when desired.

With the power of the personal computer available, it is reasonable to start with $n = 100$ subintervals. We then double the number of subintervals, run MIDPOINT again with $n = 200$ subintervals, and compare the answers. It is customary to continue this process of doubling the number of subintervals until two successive runs produce the same answer to the

```
100 REM--Program MIDPOINT
110 REM--Computes the midpoint approximation to the
120 REM--integral of the function f(x) over the input
130 REM--interval [a,b] with n subintervals.  The
140 REM--function f(x) must be edited into line 190.
150 REM
160       PRINT "EDIT IN YOUR FUNCTION F(X), THEN RUN 180"
170       EDIT 190
180       DEFDBL F, A, B, H, M, S, X
190       DEF FNF(X) =
200       INPUT "ENDPOINTS A, B"; A, B
210       PRINT
220       INPUT "NUMBER N OF SUBINTERVALS"; N
230 REM
240       H = (B - A)/N  :  S = 0  :  M = A + H/2
250       FOR I = 1 TO N
260           S = S + FNF(M)
270           M = M + H
280       NEXT I
290       S = H*S
300       PRINT USING "MIDPOINT SUM = #.#####"; S
310 REM
320       PRINT
330       INPUT "WANT ANOTHER RUN (Y/N)"; Y$
340       IF Y$ = "Y" OR Y$ = "y" THEN GOTO 210
350       END
```

Listing 3.6 Program MIDPOINT

desired number of decimal places. Hence line 330 gives us the option of making another run or terminating the program.

Example 1

Approximate the area under the parabola $y = 3x^2$ from $x = 0$ to $x = 1$.

Solution. Since the derivative of $F(x) = x^3$ is $f(x) = 3x^2$, Formula (4) tells us that the exact answer is

$$\int_0^1 3x^2\, dx = (1)^3 - (0)^3 = 1.$$

As shown in Figure 3.7, we get this value (to five decimal places) when we run Program MIDPOINT with $n = 400$.

The number π is—by definition—the area of the unit circle (radius $r = 1$), whatever this turns out to be. Looking at the unit quarter-circle shown in Figure 3.8, we see that the value of π is therefore given by the integral

$$\pi = \int_0^1 4\sqrt{1 - x^2}\, dx, \tag{11}$$

which yields four times the area of the quarter-circle.

```
RUN
EDIT IN YOUR FUNCTION F(X), THEN RUN 180
190         DEF FNF(X) = 3*X*X
RUN 180
ENDPOINTS A, B? 0, 1

NUMBER N OF SUBINTERVALS? 100
MIDPOINT SUM = 0.99997

WANT ANOTHER RUN (Y/N)? Y

NUMBER N OF SUBINTERVALS? 200
MIDPOINT SUM = 0.99999

WANT ANOTHER RUN (Y/N)? Y

NUMBER N OF SUBINTERVALS? 400
MIDPOINT SUM = 1.00000
```

Figure 3.7 Approximating the area under the parabola $y = 3x^2$ from $x = 0$ to $x = 1$

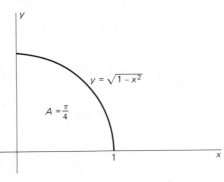

Figure 3.8 The unit quarter-circle with area $A = \pi/4$

Example 2

Approximate the integral in (11), and thus the number π.

Solution. Figure 3.9 shows the results of four successive runs of Program MIDPOINT with $f(x) = 4\sqrt{1 - x^2}$. The final two runs (with $n = 400$ and with $n = 800$) suggest that $\pi \approx 3.1416$ to four decimal places.

The final run in Figure 3.9 required about $1\frac{1}{2}$ minutes, mainly because of the time required to calculate 800 double-precision square roots. A better integral for approximating π stems from the inverse tangent function

$$\arctan (x) = \text{that angle (in radians) between } -\frac{\pi}{2} \text{ and } \frac{\pi}{2} \text{ whose tangent is } x.$$

For instance, $\arctan (1) = \pi/4$ because $\tan 45° = 1$. It is shown in Section 8.3 of Edwards and Penney, *Calculus* that the derivative of $F(x) = \arctan (x)$ is

```
RUN
EDIT IN YOUR FUNCTION F(X), THEN RUN 180
190        DEF FNF(X) = 4*SQR(1 - X*X)
RUN 180
ENDPOINTS A, B? 0, 1

NUMBER N OF SUBINTERVALS? 100
MIDPOINT SUM = 3.14194

WANT ANOTHER RUN (Y/N)? Y

NUMBER N OF SUBINTERVALS? 200
MIDPOINT SUM = 3.14171

WANT ANOTHER RUN (Y/N)? Y

NUMBER N OF SUBINTERVALS? 400
MIDPOINT SUM = 3.14164

WANT ANOTHER RUN (Y/N)? Y

NUMBER N OF SUBINTERVALS? 800
MIDPOINT SUM = 3.14161
```

Figure 3.9   Approximating the number π

$f(x) = 1/(1 + x^2)$. Therefore, Formula (4) tells us that

$$\int_0^1 \frac{dx}{1 + x^2} = \arctan(1) - \arctan(0) = \frac{\pi}{4}.$$

Consequently, we can approximate the number π by approximating the integral

$$\pi = \int_0^1 \frac{4\,dx}{1 + x^2}. \tag{12}$$

Example 3

Approximate the integral in (12) and hence the number π.

Solution.   The runs shown in Figure 3.10 indicate that only $n = 400$ intervals are required to convince us that $\pi \approx 3.14159$ accurate to five decimal places.

PROBLEMS

Use Program MIDPOINT to approximate each of the integrals in Problems 1-10. The fundamental theorem of calculus yields the exact values given in Problems 1-5. However, the integrals in Problems 6-10 are nonelementary, and therefore can only be approximated numerically.

1. $\int_2^4 x^3\,dx = 60$ **2.** $\int_1^4 \frac{dx}{\sqrt{x}} = 2$

```
RUN
EDIT IN YOUR FUNCTION F(X), THEN RUN 180
190        DEF FNF(X) = 4/(1 + X*X)
RUN 180
ENDPOINTS A, B? 0, 1

NUMBER N OF SUBINTERVALS? 100
MIDPOINT SUM = 3.14160

WANT ANOTHER RUN (Y/N)? Y

NUMBER N OF SUBINTERVALS? 200
MIDPOINT SUM = 3.14159

WANT ANOTHER RUN (Y/N)? Y

NUMBER N OF SUBINTERVALS? 400
MIDPOINT SUM = 3.14159
```

Figure 3.10   Approximating the number π

3. $\displaystyle\int_0^2 x\sqrt{1 + 2x^2}\, dx = \frac{13}{3}$

4. $\displaystyle\int_0^1 \sin^4 \pi x\, dx = \frac{3}{8}$

5. $\displaystyle\int_0^{1/2} \frac{dx}{\sqrt{1 - x^2}} = \frac{\pi}{6}$

6. $\displaystyle\int_0^1 e^{-x^2}\, dx$

7. $\displaystyle\int_0^1 \frac{dx}{\sqrt{1 + x^3}}$

8. $\displaystyle\int_1^2 \sqrt{1 + x^4}\, dx$

9. $\displaystyle\int_0^1 \frac{dx}{x^3 + x + 1}$

10. $\displaystyle\int_0^1 \frac{\sin x}{x}\, dx$

11. In Section 7.2 of Edwards and Penney, *Calculus* the natural logarithm $\log(x)$ is defined by

$$\log(x) = \int_1^x \frac{1}{t}\, dt.$$

Approximate this integral to verify that (a) $\log 2 \approx 0.69315$; (b) $\log 10 \approx 2.30259$.

12. According to the prime number theorem, the number of primes between the large positive integers a and b (with $a < b$) is given to a close approximation by the integral

$$\int_a^b \frac{dx}{\log x}.$$

Use the midpoint method to approximate the value of this integral with $a = 90,000$ and $b = 100,000$. The actual number of primes in this range is 879.

13. Alter Program MIDPOINT so that, after computing the midpoint approximation M_n with n equal subintervals, it automatically doubles the number of subintervals and computes M_{2n}—without asking whether you want another run—continuing until two successive approximations differ by less than a preassigned error tolerance.

3.2 THE TRAPEZOIDAL APPROXIMATION

One of the most frequently used numerical integration techniques employs trapezoids rather than rectangles to approximate the area under the curve $y = f(x)$ from $x = a$ to $x = b$. As indicated in Figure 3.11, the interval $[a, b]$ is subdivided into n equal subintervals of length $h = \Delta x = (b - a)/n$, by means of subdivision points $x_0 = a, x_1, x_2 \ldots, x_n = b$. On the ith subinterval $[x_{i-1}, x_i]$ a trapezoid is erected with height $f(x_{i-1})$ on the left and height $f(x_i)$ on the right. The area of this ith trapezoid is

$$\Delta A_i = \frac{h}{2}[f(x_{i-1}) + f(x_i)], \qquad (1)$$

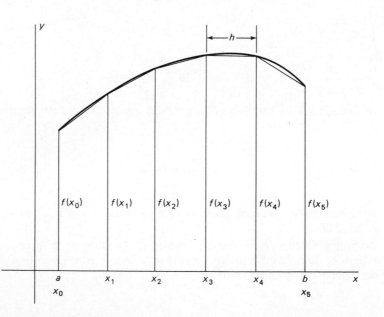

Figure 3.11 The trapezoids comprising the trapezoidal approximation (with $n = 5$ subintervals)

so the sum of the areas of the n trapezoids is

$$\sum_{i=1}^{n} \Delta A_i = \sum_{i=1}^{n} \frac{h}{2}[f(x_{i-1}) + f(x_i)]$$

$$= \frac{h}{2}[f(x_0) + f(x_1) + f(x_1) + f(x_2) + \cdots + f(x_{n-1}) + f(x_n)]$$

$$= \frac{h}{2}[f(x_0) + 2f(x_1) + 2f(x_2) + \cdots + 2f(x_{n-1}) + f(x_n)].$$

This computation yields the *trapezoidal approximation*

$$\int_{a}^{b} f(x)\, dx \approx \frac{h}{2}[f(x_0) + 2f(x_1) + \cdots + 2f(x_{n-1}) + f(x_n)], \qquad (2)$$

where $h = (b - a)/n$ and $x_i = a + ih$ for $i = 0, 1, 2, \ldots, n$. Note the
1-2-2-$\cdots$-2-1 pattern of coefficients that determines the trapezoidal
approximation.

```
100 REM--Program TRAPZOID
110 REM--Computes trapezoidal approximations to the
115 REM--integral of f(x) from x=a to x=b. The function
120 REM--f(x) is edited upon request into line 250, and
130 REM--the interval [a,b] is input at line 260. Line
140 REM--Line 290 specifies that the first approximation be
150 REM--calculated with n = 10 subintervals. A new
160 REM--approximation with n doubled is calculated until
170 REM--two successive approximations agree to within the
180 REM--error tolerance specified in line 420.
190 REM
200 REM--Initialization:
210 REM
220       PRINT "EDIT IN YOUR FUNCTION F(X), THEN RUN 240"
230       EDIT 250
240       DEFDBL A,B,F,H,T,X  :  DEFINT I,K,N
250       DEF FNF(X) =
260       INPUT "ENDPOINTS A, B"; A, B
270       PRINT : PRINT "Number of",     "Trapezoidal"
280            PRINT "Subintervals", "Approximation" :
                 PRINT
290       N = 10    :    TOLD = 0
300 REM
310 REM--Trapezoidal sum loop:
320 REM
330       X = A  :  H = (B - A)/N  :  TNEW = 0
340       FOR I=0 TO N
350            IF  I=0 OR I=N  THEN  K=1  ELSE  K=2
360            TNEW = TNEW + K*FNF(X)
370            X = X + H
380       NEXT I
```

Listing 3.12 Program TRAPZOID

```
390        TNEW = (H/2)*TNEW
400        PRINT N, :  PRINT USING "#.######"; TNEW
410 REM
420        IF ABS(TNEW - TOLD) < .000003 THEN GOTO 450
430        N = 2*N  :  TOLD = TNEW  :  GOTO 330
440 REM
450        PRINT  :
           PRINT USING "INTEGRAL  = #.#####"; TNEW
460        END
```

Listing 3.12 (con't.)

Listing 3.12 exhibits Program TRAPZOID, which computes the approximation in (2). The function $f(x)$ is (upon request at line 230) edited into line 250 when the program is run, and the interval $[a, b]$ is input at line 260. The first trapezoidal approximation is computed with $n = 10$ subintervals (as specified in line 290). The running sum

$$\text{TNEW} = f(x_0) + 2f(x_1) + 2f(x_2) + \cdots$$

is calculated in the FOR-NEXT loop consisting of lines 340–380, and the complete sum is finally multiplied by $h/2$ in line 390. Note that the characteristic 1-2-2-$\cdots$-2-1 pattern of coefficients appears in line 350. Lines 420 and 430 direct that n be doubled and a new trapezoidal approximation computed, with this process being repeated until two successive approximations differ by less than the error tolerance specified in line 420, at which point the apparent value of the integral is printed.

Example 1

Approximate the integral

$$\pi = \int_0^1 \frac{4\,dx}{1 + x^2}.$$

Solution. The run of Program TRAPZOID shown in Figure 3.13 yields $\pi \approx 3.14159$.

Example 2

Approximate the integral

$$\log 2 = \int_1^2 \frac{dx}{x}.$$

Solution. The run of Program TRAPZOID shown in Figure 3.14 yields the result $\log 2 \approx 0.69315$, which is the correct value rounded off to five decimal places.

```
RUN
EDIT IN YOUR FUNCTION F(X), THEN RUN 240
250       DEF FNF(X) = 4/(1 + X*X)
RUN 240
ENDPOINTS A, B? O, 1

Number of          Trapezoidal
Subintervals       Approximation

10                 3.139926
20                 3.141176
40                 3.141488
80                 3.141567
160                3.141586
320                3.141591
640                3.141592

INTEGRAL  = 3.14159
```

Figure 3.13   Approximating π with TRAPZOID

```
RUN
EDIT IN YOUR FUNCTION F(X), THEN RUN 240
250       DEF FNF(X) = 1/X
RUN 240
ENDPOINTS A, B? 1, 2

Number of          Trapezoidal
Subintervals       Approximation

10                  0.693771
20                  0.693303
40                  0.693186
80                  0.693157
160                 0.693150
320                 0.693148

INTEGRAL  = 0.69315
```

Figure 3.14   Approximating log 2 with TRAPZOID

Improving the Accuracy

As soon as two successive trapezoidal approximations agree to within the prescribed error tolerance, Program TRAPZOID terminates the iteration and prints "INTEGRAL = (whatever)." Of course, the computer prints this not because it is necessarily true, but merely because we programmed it to do so. Hence the question remains as to how accurate our "answer" is.

To discuss the accuracy of the trapezoidal approximation to the integral

$$I = \int_a^b f(x)\ dx, \tag{3}$$

let us write $T(h)$ for the trapezoidal approximation with a fixed $h = (b - a)/n$, that is,

$$T(h) = \frac{h}{2}[f(x_0) + 2f(x_1) + \cdots + 2f(x_{n-1}) + f(x_n)]. \qquad (4)$$

In numerical analysis texts it is proved that given n and h, there exists a point c between a and b such that

$$\int_a^b f(x)\, dx = T(h) - \frac{b - a}{12} f''(c)h^2. \qquad (5)$$

If M denotes the maximum value of $|f''(x)|$ on the interval $[a, b]$, it follows that the *error*

$$E(h) = \int_a^b f(x)\, dx - T(h) \qquad (6)$$

in the trapezoidal approximation $T(h)$ is bounded by

$$|E(h)| \leqq \frac{M}{12}(b - a)h^2. \qquad (7)$$

Consequently, if h is "very small," then the error is "very, very small," because h^2 is small compared with h.

Although Inequality (7) implies that $E(h)$ approaches 0 very rapidly as $h \longrightarrow 0$, it yields only an approximate knowledge of $E(h)$; the actual error generally is somewhat smaller than the upper bound in (7). However, this is a fortunate situation in which we can "use an approximate knowledge of the error to approximately eliminate this error," and thereby get a much more accurate estimate of the integral I.

To do this, let us write

$$T_{n-1} = T(h) \quad \text{and} \quad T_n = T\left(\frac{h}{2}\right) \qquad (8)$$

to denote two successive trapezoidal approximations to the integral I, the latter with twice as many subintervals as the former. Then Formula (5) yields

$$\int_a^b f(x)\, dx = T_{n-1} - \frac{1}{12}(b - a)f''(c_1)h^2 \qquad (9a)$$

and

$$\int_a^b f(x)\ dx = T_n - \frac{1}{12}(b-a)f''(c_2)\left(\frac{h}{2}\right)^2 \qquad (9b)$$

for appropriate points c_1 and c_2 in (a, b).

It seems plausible that $f''(c_1) \approx f''(c_2)$ if all is well, so let us write

$$K \approx \frac{1}{12}(b-a)f''(c_1) \approx \frac{1}{12}(b-a)f''(c_2) \qquad (10)$$

for the approximate common value. Then from (9a) and (9b) we get

$$I \approx T_{n-1} - Kh^2, \qquad (11a)$$

$$I \approx T_n - \frac{1}{4}Kh^2. \qquad (11b)$$

If we rewrite these two approximate equalities as

$$4I \approx 4T_n - Kh^2,$$

$$I \approx T_{n-1} - Kh^2,$$

subtract, and divide by 3, then we get

$$I \approx \frac{1}{3}(4T_n - T_{n-1}). \qquad (12)$$

In short, the fact that the error in T_n is only about one-fourth of the error in T_{n-1} has led us to discover the propitious combination

$$R_n = \frac{1}{3}(4T_n - T_{n-1}) \qquad (13)$$

of successive trapezoidal sums. Because R_n would *exactly* equal the integral I if the error in T_n were *precisely* one-fourth of the error in T_{n-1}, we may reasonably expect R_n to be a considerably more accurate approximation than either T_n or T_{n-1}.

Combining two successive trapezoidal approximations T_{n-1} and T_n as in (13) to obtain a more accurate approximation R_n is known as *Richardson extrapolation*. If $T_1, T_2, T_3, \ldots$ is a sequence of trapezoidal approximations—with each computed using twice as many subintervals as the preceding one—then the sequence of successive extrapolations $R_2, R_3, R_4, \ldots$ should exhibit much more rapid convergence to the value of the integral.

```
100 REM--Program EXTRAP
110 REM--An alteration of Program TRAPZOID that computes
115 REM--both trapezoidal approximations and Richardson
120 REM--extrapolations to the value of the integral of
130 REM--the function f(x) from x=a to x=b.
140 REM
200 REM--Initialization:
210 REM
220        PRINT "EDIT IN YOUR FUNCTION F(X), THEN RUN 240"
230        EDIT 250
240        DEFDBL A,B,F,H,T,X  :  DEFINT I,K,N
250        DEF FNF(X) =  1/X
260        INPUT "ENDPOINTS A, B"; A, B
270        PRINT  :  PRINT
           "Number of",     "Trapezoidal"
280        PRINT
           "Subintervals", "Approximation", "Extrapolation"
285        PRINT
290        N = 10   :    TOLD = 0
300 REM
310 REM--Trapezoidal sum loop:
320 REM
330        X = A  :  H = (B - A)/N  :   TNEW = 0
340        FOR I=0 TO N
350             IF  I=0 OR I=N  THEN  K=1  ELSE  K=2
360             TNEW = TNEW + K*FNF(X)
370             X = X + H
380        NEXT I
390        TNEW = (H/2)*TNEW
400        PRINT N, :  PRINT USING "#.#####"; TNEW
405        IF N > 10 THEN PRINT " ", " ",  :
           PRINT USING "  #.##########"; (4*TNEW - TOLD)/3
410 REM
420        IF ABS(TNEW - TOLD) < .0001 THEN GOTO 450
430        N = 2*N  :   TOLD = TNEW  :   GOTO 330
440 REM
450        PRINT  :  PRINT USING
           "INTEGRAL  =   #.##########"; (4*TNEW - TOLD)/3
460        END
```

Listing 3.15   Program EXTRAP

Listing 3.15 shows Program EXTRAP, which is a very slight alteration of Program TRAPZOID. Aside from remarks and format instructions, it differs only in that we have added line 405 to compute and print at each stage the extrapolation $(4 * TNEW - TOLD)/3$ from the preceding two trapezoidal approximations. Figure 3.16 shows a run of Program EXTRAP for the familiar integral

$$\pi = \int_0^1 \frac{4\,dx}{1 + x^2}.$$

Observe that with only $n = 40$ subintervals the extrapolated value gives π accurate to 10 decimal places (rounded off), whereas the corresponding trapezoidal approximation is accurate to only three decimal places.

```
RUN
EDIT IN YOUR FUNCTION F(X), THEN RUN 240
250      DEF FNF(X) = 4/(1 + X*X)
RUN 240
ENDPOINTS A, B? 0, 1
```

Number of Subintervals	Trapezoidal Approximation	Extrapolation
10	3.13993	
20	3.14118	
		3.1415926530
40	3.14149	
		3.1415926536
80	3.14157	
		3.1415926536

```
INTEGRAL  =  3.1415926536
```

Figure 3.16   Approximating π by extrapolation

PROBLEMS

1-10. Run Program TRAPZOID to approximate the integral of the corresponding problem in the Section 3.1 problems.

11. Use Program EXTRAP to compute log 2, as given by the integral of Example 2, accurate to 10 decimal places.

12. Consider the ellipse $x^2/a^2 + y^2/b^2 = 1$ with eccentricity $\epsilon = (a^2 - b^2)^{1/2}/a$. Its perimeter is given by the the integral

$$p = \int_0^{\pi/2} 4a\sqrt{1 - \epsilon^2 \sin^2\theta}\ d\theta.$$

Approximate the value of this integral with $a = 10$ and $\epsilon = 1/2$.

13. The *approximate* formula $T = 2\pi(L/g)^{1/2}$, for the period of oscillation of a pendulum of length L, yields $T = \pi/2 \approx 1.5708$ sec if $L = 2$ ft and $g = 32$ ft/sec^2. If this same pendulum is released from rest with an initial angular displacement of 60°, its *actual* period is

$$T = \int_0^{\pi/2} \frac{2\,d\theta}{\sqrt{4 - \sin^2\theta}}.$$

Approximate this integral.

14. The trapezoidal approximation with endpoint correction is

$$\int_a^b f(x)\,dx \approx T(h) - \frac{h^2}{12}[f'(b) - f'(a)].$$

(a) Alter Program TRAPZOID so that it incorporates the endpoint correction, assuming that the values $f'(a)$ and $f'(b)$ are input initially.

(b) Approximate $\log 2$ as in Example 2 by running this altered program with $a = 1$, $b = 2$, $f(x) = 1/x$, and $f'(x) = -1/x^2$.

15. The number e may be defined as that number such that $\log e = 1$. That is, e is the solution of the equation

$$f(x) = \left(\int_1^x \frac{dt}{t} \right) - 1 = 0.$$

Noting that $f'(x) = 1/x$ by the fundamental theorem of calculus, write a program that applies Newton's method to solve this equation for e. Use a trapezoidal approximation subroutine to calculate values of $f(x)$.

3.3 SIMPSON'S APPROXIMATION

The Richardson extrapolation discussed in Section 3.2 provides an optimal way of combining two trapezoidal approximations T_0 and T_1, with T_1 employing twice as many subintervals as T_0. Let the interval $[a, b]$ of integration be subdivided into an *even* number $2n$ of equal subintervals by means of the points $x_0, x_1, x_2, \ldots, x_{2n-1}, x_{2n}$. However, when we form the first trapezoidal approximation we consider only the points $x_0, x_2, \ldots, x_{2n-2}, x_{2n}$, so the ith subinterval $[x_{2i-2}, x_{2i}]$ has length $2h$, where

$$h = \frac{b - a}{2n}. \tag{1}$$

Consequently, our first approximation to $\int_a^b f(x)\, dx$, with n subintervals, is

$$T_0 = \frac{2h}{2}[f(x_0) + 2f(x_2) + \cdots + 2f(x_{2n-2}) + f(x_{2n})]. \tag{2}$$

Our second trapezoidal approximation, with $2n$ subintervals of length h, is

$$T = \frac{h}{2}[f(x_0) + 2f(x_1) + 2f(x_2) + \cdots + 2f(x_{2n-1}) + f(x_{2n})]. \tag{3}$$

We now consider the Richardson extrapolation

$$\frac{1}{3}(4T_1 - T_0) = \frac{4}{3}T_1 - \frac{1}{3}T_0$$

$$= \frac{2h}{3}[f(x_0) + 2f(x_1) + 2f(x_2) + \cdots + 2f(x_{2n-1}) + f(x_{2n})]$$

$$- \frac{h}{3}[f(x_0) + 2f(x_2) + \cdots + 2f(x_{2n-2}) + f(x_{2n})]$$

$$= \frac{h}{3}[f(x_0) + 4f(x_1) + 2f(x_2) + \cdots + 2f(x_{2n-2})$$
$$+ 4f(x_{2n-1}) + f(x_{2n})].$$

The latter sum is probably the most commonly used of all numerical approximations to the integral

$$I = \int_a^b f(x) \, dx.$$

It is *Simpson's approximation*,

$$S_{2n} = \frac{h}{3}(y_0 + 4y_1 + 2y_2 + 4y_3 + \cdots + 2y_{2n-2} + 4y_{2n-1} + y_{2n}), \qquad (4)$$

where $y_i = f(y_i)$ and $h = (b - a)/2n$. The subscript $2n$ indicates that $2n$ subintervals are involved.

Remark: Observe the 1-4-2-4-$\cdots$-2-4-1 pattern of coefficients in Simpson's approximation. This pattern "works out" only with an *even* number $2n$ of subintervals (in contrast with the trapezoidal approximation, for which the number of subintervals may be either even or odd).

Formula (4) for Simpson's approximation is often derived in other ways. For instance, the reader may verify that

$$S_{2n} = \frac{2}{3}M_n + \frac{1}{3}T_n, \qquad (5)$$

where M_n and T_n denote the midpoint approximation and the trapezoidal approximation, respectively, each with n subintervals (rather than $2n$ subintervals).

Another approach involves approximating the curve $y = f(x)$ with parabolic arcs. On the ith subinterval $[x_{2i-2}, x_{2i}]$ the curve $y = f(x)$ is replaced with the parabolic arc

$$y = p_i(x) = A_i x^2 + B_i x + C_i,$$

where the coefficients A_i, B_i, and C_i are chosen so that $f(x)$ and $p_i(x)$ agree at the three points x_{2i-2}, x_{2i-1}, and x_{2i}, that is, so that

$$f(x_{2i-2}) = p_i(x_{2i-2}),$$
$$f(x_{2i-1}) = p_i(x_{2i-1}),$$
$$f(x_{2i}) = p_i(x_{2i}).$$

Then it turns out that the approximation

$$\int_a^b f(x)\ dx \approx \sum_{i=1}^{n} \int_{x_{2i-2}}^{x_{2i}} p_i(x)\ dx$$

reduces to Simpson's approximation (4). See Section 5.8 of Edwards and Penney, *Calculus* for additional details.

```
100 REM--Program SIMPSON
110 REM--Computes Simpson's approximations to the integral
115 REM--of f(x) from x=a to x=b.  The function f(x) is
120 REM--edited upon request into line 250, and the
130 REM--interval [a,b] is input at line 260.  Line 290
140 REM--specifies that the first approximation be
150 REM--calculated with n = 10 subintervals.  A new
160 REM--approximation with n doubled is calculated until
170 REM--two successive approximations agree to within the
180 REM--error toleramce specified in line 420.
190 REM
200 REM--Initialization:
210 REM
220        PRINT "EDIT IN YOUR FUNCTION F(X), THEN RUN 240"
230        EDIT 250
240        DEFDBL A,B,F,H,S,X  :   DEFINT I,K,N
250        DEF FNF(X) =
260        INPUT "ENDPOINTS A, B"; A, B
270        PRINT : PRINT "Number of",     "Simpson's"
280              PRINT "Subintervals", "Approximation" :
                 PRINT
290        N = 10    :    SOLD = 0
300 REM
310 REM--Simpson's rule loop:
320 REM
330        X = A  :  H = (B - A)/N  :  SNEW = 0
340        FOR I=0 TO N
350            IF  I=0 OR I=N   THEN   K=1  ELSE
                   IF I/2 = I\2 THEN   K=2  ELSE   K=4
360              SNEW = SNEW + K*FNF(X)
370              X = X + H
380        NEXT I
390        SNEW = (H/3)*SNEW
400        PRINT N, :  PRINT USING "#.##########"; SNEW
410 REM
420        IF ABS(SNEW - SOLD) < 1E-11 THEN GOTO 450
430        N = 2*N  :   SOLD = SNEW  :    GOTO 330
440 REM
450        PRINT  :  PRINT USING
           "INTEGRAL  = #.##########"; SNEW
460        END
```

Listing 3.17   Program SIMPSON

Listing 3.17 shows Program SIMPSON. Observe that—except for remarks and print statements—it is essentially indentical with Program TRAPZOID (Listing 3.12). Line 350 has been altered in accordance with the 1-4-2- · · · -2-4-1 pattern of coefficients in Simpson's approximation, and $h/3$ instead of $h/2$ appears in line 390. Indeed, we wrote SIMPSON simply by loading Program TRAPZOID and editing it slightly. It is good programming practice to rely as much as possible on alteration (and cannibalization) of existing programs.

Note that in line 350 we use "integer division" to sort out the cases. The symbol $I\backslash 2$ denotes the integer part of the quotient $I/2$. For instance, $5\backslash 2 = \text{Int}(5/2) = 2 \neq 5/2$ while $6\backslash 2 = \text{Int}(6/2) = 3 = 6/2$. Thus $I/2 = I\backslash 2$ if and only if the integer I is even.

```
RUN
EDIT IN YOUR FUNCTION F(X), THEN RUN 240
250        DEF FNF(X) =   4/(1 + X*X)
RUN 240
ENDPOINTS A, B? O, 1

Number of        Simpson's
Subintervals     Approximation

10               3.14159261394
20               3.14159265297
40               3.14159265358
80               3.14159265359

INTEGRAL  = 3.1415926536
```

Figure 3.18   Approximating π with Simpson's rule

Figure 3.18 shows a run of Program SIMPSON to approximate our old friend

$$\pi = \int_0^1 \frac{4\,dx}{1 + x^2}.$$

With only 80 subintervals we get the correct value of π rounded off to 10 decimal places.

Improper Integrals

To be assured that the numerical approximations we have discussed converge to the actual value of the integral $\int_a^b f(x)\,dx$ as $n \longrightarrow \infty$ ($n =$ number of subintervals), we need to know that the interval $[a, b]$ is finite and also that the values of $f(x)$ are bounded on $[a, b]$. Otherwise, the integral is called *improper*. For instance, the integral

$$\int_0^1 \frac{dx}{\sqrt{x}} \tag{6}$$

is improper because $1/\sqrt{x} \longrightarrow \infty$ as $x \longrightarrow 0$. The meaning of this improper integral is

$$\int_0^1 \frac{dx}{\sqrt{x}} = \lim_{\epsilon \to 0} \int_\epsilon^1 \frac{dx}{\sqrt{x}} \tag{7}$$

provided that this limit exists; note that if $\epsilon > 0$, the integral on the right is proper. Indeed,

$$\int_0^1 \frac{dx}{\sqrt{x}} = \lim_{\epsilon \to 0} \int_\epsilon^1 \frac{dx}{\sqrt{x}}$$

$$= \lim_{\epsilon \to 0} [2\sqrt{x}]_\epsilon^1$$

$$= \lim_{\epsilon \to 0} (2\sqrt{1} - 2\sqrt{\epsilon}) = 2. \tag{8}$$

We defer numerical approximation of improper integrals like (6) to the problems, and concentrate here on improper integrals like

$$\int_0^\infty f(x)\, dx, \tag{9}$$

in which the function is bounded but the interval $[0, \infty)$ of integration is infinite. For instance, the integral

$$\int_0^\infty e^{-x^2}\, dx \tag{10}$$

is important in probability and statistics (and elsewhere). Its meaning is given by

$$\int_0^\infty e^{-x^2}\, dx = \lim_{b \to \infty} \int_0^b e^{-x^2}\, dx. \tag{11}$$

However, the function $e^{-x^2}\, dx$ has no elementary antiderivative as a finite combination of familiar functions, so a simple and direct evaluation of the limit in (11) using only the fundamental theorem of calculus is not feasible. The exact value is known—by indirect methods—to be

$$\int_0^\infty e^{-x^2}\, dx = \frac{\sqrt{\pi}}{2} \approx 0.886226926. \tag{12}$$

Figure 3.19 illustrates our approach to a numerical verification of the limit in (12) and, more generally, to the numerical approximation of improper integrals of the form in (9). We use Simpson's rule first to approximate the area under $y = e^{-x^2}$ from $x = 0$ to $x = 1$, next to approximate the area from $x = 1$ to $x = 10$, then the area from $x = 10$ to $x = 100$, and so on. We continue to add these areas to our running sum until we reach the point that the area from $x = 10^k$ to $x = 10^{k+1}$ is less than our preassigned error tolerance.

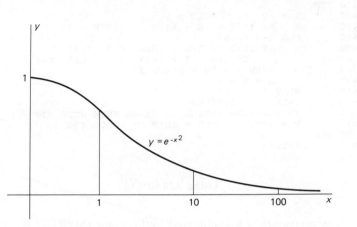

Figure 3.19 The area under $y = e^{-x^2}$ (not to scale)

```
100 REM--Program IMPROPER
110 REM--Uses Simpson's rule to approximate the improper
115 REM--integral from  x = 0  to  x = infinity  of the
120 REM--function f(x) that is edited into line 250.
130 REM--First the integral from  x = 0 to x = 1 is
140 REM--calculated, then the integral  from  x = 1 to
150 REM--x = 10, then the integral from  x = 10   to
160 REM--x = 100, and so forth, continuing until the
165 REM--contribution of the new integral is less than
170 REM--the error tolerance specified in line 350.
180 REM
200 REM--Initialization:
210 REM
220       PRINT "EDIT IN YOUR FUNCTION F(X), THEN RUN 240"
230       EDIT 250
240       DEFDBL A,B,F,H,S,T,X    :    DEFINT I,K,N
250       DEF FNF(X) =
260       PRINT : PRINT "Upper",    "Simpson's"
270             PRINT "limit",   "approximation"  :
                PRINT
280       A = 0   :    B = 1   :    TOLD = 0
290 REM
300 REM--Main program:
310 REM
320       GOSUB 400  :   REM--Integrate over new subinterval
330       TNEW = TOLD + SNEW
340       PRINT B,  :   PRINT USING "#.########"; TNEW
350       IF ABS(SNEW) < 1E-08 THEN GOTO 380
360       A = B   :    B = 10*A   :    REM--New limits
370       TOLD = TNEW   :    GOTO 320
380       PRINT  :   PRINT USING
                "INTEGRAL  =   #.#######"; TNEW
390       STOP
395 REM
```

Listing 3.20 Program IMPROPER

```
400 REM--Simpson's rule subroutine:
410 REM
420      N = 10   :    SOLD = O
430      X = A  :   H = (B - A)/N  :   SNEW = O
440      FOR I=O TO N
450           IF  I=O OR I=N   THEN   K=1   ELSE
                  IF I/2 = I\2 THEN   K=2   ELSE   K=4
460           SNEW = SNEW + K*FNF(X)
470           X = X + H
480      NEXT I
490      SNEW = (H/3)*SNEW
500      IF ABS(SNEW - SOLD) < 1E-08 THEN GOTO 530
510      N = 2*N  :   SOLD = SNEW  :   GOTO 430
520 REM
530      RETURN
540      END
```

Listing 3.20 (con't.)

This procedure is implemented by Program IMPROPER shown in Listing 3.20. The variable TNEW denotes the running sum

$$\int_0^{10^k} f(x) \, dx = \int_0^1 f(x) \, dx + \sum_{i=1}^{k} \int_{10^{i-1}}^{10^i} f(x) \, dx. \tag{13}$$

Each new integral

$$\text{SNEW} \approx \int_{10^{k-1}}^{10^k} f(x) \, dx \tag{14}$$

on the right-hand side in (13) is approximated using the Simpson's rule subroutine in lines 400–530.

```
RUN
EDIT IN YOUR FUNCTION F(X), THEN RUN 240
250        DEF FNF(X) =   EXP(-X*X)
RUN 240

Upper              Simpson's
Limit              Approximation

1                  0.74682413
10                 0.88622693
100                0.88622693

INTEGRAL    =      0.8862269
```

Figure 3.21 Approximating the improper integral $\int_0^\infty e^{-x^2} \, dx$

Figure 3.21 shows the output of Program IMPROPER with $f(x) = e^{-x^2}$. Note that with a rather conservative error tolerance of 10^{-8} specified in lines 350 and 500, the integral from $x = 0$ to $x = 10$ agrees with the exact result in (12) rounded off to seven decimal places.

To approximate an improper integral $\int_a^b f(x)\,dx$ with lower limit $a > 0$, you should assign the initial values A $= a$ and B $= 10a$ in line 280. Alternatively, you can replace line 280 with the lines

$$280 \quad \text{INPUT "LOWER LIMIT A"; A}$$
$$285 \quad \text{B} = 10 * \text{A : TOLD} = 0$$

so as to input A during execution of the program.

PROBLEMS

1-10. Run Program SIMPSON to approximate the integral of the corresponding problem in the Section 3.1 Problems.

11. Use Program SIMPSON to find ln 2 and ln 10 accurate to 10 decimal places.

12. Assume that the moon's orbit is an ellipse with the earth at one focus, with its major semiaxis being exactly $a = 238,857$ miles and its eccentricity being exactly $\epsilon = 0.0549$. Referring to Problem 12 in Section 3.2, use Program SIMPSON to find the perimeter of the moon's orbit, accurate to the nearest foot.

Use Program IMPROPER to approximate the integrals in Problems 13-18.

13. $\displaystyle\int_0^\infty e^{-x}\,dx = 1$ **14.** $\displaystyle\int_0^\infty \frac{dx}{x^2} = 1$

15. $\displaystyle\int_1^\infty \frac{dx}{x^2 + 1} = \frac{\pi}{4}$ **16.** $\displaystyle\int_1^\infty \frac{\sin x}{x}\,dx$

17. $\displaystyle\int_0^\infty \frac{dx}{\sqrt{x^3 + 1}}$ **18.** $\displaystyle\int_1^\infty \frac{\ln x}{x^2}\,dx$

19. Suppose that $\int_0^1 f(x)\,dx$ is—like integral (6) in this section—improper only at $x = 0$. Write a program to approximate the value of this integral, as follows. Use Simpson's rule first to approximate the area under $y = f(x)$ from $x = 0.1$ to $x = 1$, next to approximate the area from $x = 0.01$ to $x = 0.1$, then the area from $x = 0.001$ to $x = 0.01$, and so on.

Use the program of Problem 19 to approximate the integrals in Proglems 20-24.

20. $\displaystyle\int_0^1 \frac{dx}{\sqrt{x}} = 2$ **21.** $\displaystyle\int_0^1 \frac{dx}{x^{2/5}} = \frac{5}{3}$

22. $\displaystyle\int_0^1 \frac{e^x - 1}{x}\,dx$ **23.** $\displaystyle\int_0^1 \left(\ln \frac{1}{x}\right)^3 dx$

24. $\displaystyle\int_0^1 \ln\left(\sin \frac{\pi x}{2}\right) dx$

Verify numerically the values specified in Problems 25–29.

25. $\int_0^\infty x^5 e^{-x} \, dx = 120$

26. $\int_0^\infty \frac{\sin x}{x} \, d = \frac{\pi}{2}$

27. $\int_0^\infty \frac{dx}{x^2 + 2} = \frac{\pi}{2\sqrt{2}}$

28. $\int_0^\infty e^{-x^2} \cos 2x \, dx = \frac{\sqrt{\pi}}{2e}$

29. $\int_0^\infty \frac{1 - e^{-3x}}{x} \, dx = \frac{1}{2} \ln 10$

Infinite Series and Transcendental Functions

4

4.1 INTRODUCTION

Numerical values of the familiar transcendental functions (exponentials, logarithms, sines, cosines, etc.) are generally calculated most readily by means of infinite series. For example, the definition of the sine of an angle (in a right triangle) as the ratio of the opposite side to the hypotenuse affords no immediate and easy way of calculating the value sin 1°. However, as we shall see in this chapter, the infinite series

$$\sin x = x - \frac{x^3}{3!} + \frac{x^5}{5!} - \frac{x^7}{7!} + \cdots$$

makes this a simple matter.

Figure 4.1 An infinite subdivision of a segment of length 2

For a tangible example of an infinite series, look at the segment of length 2 in Figure 4.1. It has been subdivided first into two equal segments of length 1. Then the right segment has been subdivided into two equal segments of length 1/2, and so on *ad infinitum* (in the sense of the fleas that have little fleas that bite'm, which in turn have littler fleas that bite'm, and so on ad infinitum). This subdivision makes it appear evident that

$$2 = 1 + \frac{1}{2} + \frac{1}{4} + \cdots + \frac{1}{2^n} + \cdots. \tag{1}$$

The right-hand side here is an "infinite series" in that the additional terms continue indefinitely; each new term is one-half of its predecessor.

But precisely what is meant by saying that the infinite series in (1) adds up to the value 2? More generally, we say that the infinite series

$$a_0 + a_1 + a_2 + \cdots + a_n + \cdots \tag{2}$$

converges and has *sum S* provided that

$$\lim_{n \to \infty} (a_0 + a_1 + a_2 + \cdots + a_n) = S, \tag{3}$$

in which case we write

$$\sum_{n=0}^{\infty} a_n = S. \tag{4}$$

Equivalently, if we first form for each $n = 0, 1, 2, \ldots$ the *partial sum*

$$s_n = a_0 + a_1 + a_2 + \cdots + a_n, \tag{5}$$

then $s_0, s_1, s_2, \ldots$ is a sequence like those we discussed in Section 2.1, and (4) means that

$$\lim_{n \to \infty} s_n = S. \tag{6}$$

For the particular series in (1) we have

$$s_0 = 1, \quad s_1 = \frac{3}{2}, \quad s_2 = \frac{7}{4}, \quad s_3 = \frac{15}{8}, \cdots.$$

In general, $s_n = 2 - (1/2^n)$, so the sum of the series is

$$S = \lim_{n \to \infty} s_n = \lim_{n \to \infty} \left(2 - \frac{1}{2^n}\right) = 2,$$

as claimed.

Thus the sum of an infinite series is, in principle, a special case of the limit of an infinite sequence. However, some of the characteristic features of the summation of infinite series make it profitable to discuss the subject separately.

For an example of an infinite series whose sum is not so evident, consider the series

$$1 + \frac{3}{7} + \frac{9}{49} + \frac{27}{243} + \cdots + \left(\frac{3}{7}\right)^n + \cdots. \tag{7}$$

Both (2) and (7) are examples of the *geometric series*

$$\sum_{n=0}^{\infty} ar^n = a + ar + ar^2 + \cdots \tag{8}$$

with initial term a and ratio r—that is, r is the ratio of each term to its predecessor. With $a = 1$ and $r = 1/2$, series (8) reduces to (2), while with $a = 1$ and $r = 3/7$ it reduces to (7).

```
100 REM--Program GEOMETRC
110 REM--Computes the sum of a geometric series with
120 REM--input initial term A and ratio R.
130 REM
140       INPUT "INITIAL TERM"; A
150       INPUT "RATIO"; R
160       TERM = A    :    SUM = A
170 REM
180       PRINT SUM
190       TERM = R*TERM
200       SUM = SUM + TERM
210       IF ABS(TERM) > .00001 THEN GOTO 180
220 REM
230       PRINT USING "SUM = #.####"; SUM
240       END
```

Listing 4.2   Program GEOMETRC

The fact that each term of the geometric series (8) is r times the preceding term makes it simple to write a program to sum the series (assuming that it converges). Program GEOMETRC (Listing 4.2) first calls for us to input the initial term A and ratio R. The current term TERM and the running partial sum SUM are both initialized to A in line 160. Successive partial sums are computed and printed in the loop consisting of lines 180–210. The next term R∗TERM is calculated in line 190 and then added to SUM in line 200. The loop is exited (and the value of the SUM printed) when two successive partial sums agree to within the error tolerance $E = 0.00001$ specified in line 210.

Figure 4.3 shows two runs of Program GEOMETRC with A = 1 and with the ratios R = 0.5 = 1/2 and R = 0.4285714 ≈ 3/7 corresponding to

```
RUN
INITIAL TERM?   1
RATIO?   .5
 1
 1.5
 1.75
 1.875
 1.9375
 1.96875
 1.984375
 1.992188
 1.996094
 1.998047
 1.999024
 1.999512
 1.999756
 1.999878
 1.999939
 1.99997
 1.999985
SUM = 2.0000


RUN
INITIAL TERM?   1
RATIO?   .4285714
 1
 1.428572
 1.612245
 1.690962
 1.724698
 1.739156
 1.745353
 1.748008
 1.749147
 1.749634
 1.749843
 1.749933
 1.749971
 1.749988
SUM = 1.7500
```

Figure 4.3   Two runs of Program GEOMETRC

series (2) and (7), respectively. We can check the indicated sum 1.7500 of series (7) as follows. In Section 12.3 of Edwards and Penney, *Calculus* it is proved that the geometric series (8) converges if and only if $|r| < 1$, in which case

$$\sum_{n=0}^{\infty} ar^n = \frac{a}{1-r}. \tag{9}$$

With $a = 1$ and $r = 3/7$ we get

$$\sum_{n=0}^{\infty} \left(\frac{3}{7}\right)^n = \frac{1}{1-(3/7)} = \frac{1}{4/7} = \frac{7}{4} = 1.75.$$

As in the similar situation for limits of sequences, the stopping condition $|\text{new sum} - \text{old sum}| < E$ is a reliable convergence test only when one knows in advance that the infinite series converges fairly rapidly. For instance, the *harmonic series*

$$\sum_{n=1}^{\infty} \frac{1}{n} = 1 + \frac{1}{2} + \frac{1}{3} + \frac{1}{4} + \cdots \tag{10}$$

diverges to $+\infty$—that is, its partial sums $\{s_n\}$ approach infinity as $n \longrightarrow \infty$ (see Section 12.3 of Edwards and Penney, *Calculus*). But obviously its nth term $1/n$ is eventually smaller than any preassigned error tolerance E. Thus the simple program

```
10 N = 1 : S = 1
20 PRINT N, S
30 N = N + 1
40 S = S + 1/N
50 IF 1/N > 0.001 THEN GOTO 20
60 PRINT "SUM =    "; S
70 END
```

yields the false result SUM = 7.485479.

A more exotic example is the series

$$\sum_{n=2}^{\infty} \frac{1}{n(\ln n)^2} = \frac{1}{2(\ln 2)^2} + \frac{1}{3(\ln 3)^2} + \cdots, \tag{11}$$

which is discussed in Section 12.4 of Edwards and Penney, *Calculus*. There it is shown that this series converges, but that approximately 2.7×10^{43} terms would have to be added to find its accuracy to within E = 0.01! Thus the summation of series is like the solution of equations, in that no practical method can be expected to succeed in every case.

The Binomial Series

One of the most important infinite series in mathematics is the *binomial series*

$$(1 + x)^k = 1 + \frac{k}{1}x + \frac{k(k-1)}{1\cdot 2}x^2$$

$$+ \cdots + \frac{k(k-1)\cdots(k-n+1)}{1\cdot 2\cdot 3 \cdots n}x^n + \cdots, \tag{12}$$

which (for all real k) converges if $|x| < 1$.

If k is a positive integer then (12) reduces to a (finite) binomial formula, such as

$$(1 + x)^4 = 1 + 4x + 6x^2 + 4x^3 + x^4.$$

Otherwise, it is a genuine infinite series. For instance, with $k = 1/2$ it yields

$$\sqrt{1 + x} = 1 + \frac{1}{2}x - \frac{1}{8}x^2 + \frac{1}{16}x^3 - \frac{5}{128}x^4 + \cdots.$$

The x^{n+1} term in the binomial series is obtained by multiplying the x^n term by $x(k - n)/(n + 1)$. This explains line 200 in Program BINOMIAL (Listing 4.4), which otherwise is essentially the same as Program GEOMETRC. When the variable x and the exponent k are input, the program computes $(1 + x)^k$ to within the error tolerance specified in line 230.

```
100 REM--Program BINOMIAL
110 REM--Computes sum of the binomial series for  (1 + x)^k
120 REM--when the variable x and the exponent k are input.
130 REM
140     DEFDBL K, N, S, T, X
150     INPUT "VARIABLE X"; X
160     INPUT "EXPONENT K"; K
170     TERM = 1  :  SUM = 1  :  N = 0
180 REM
190     PRINT SUM
200     TERM = TERM*X*(K - N)/(N + 1)
210     SUM = SUM + TERM
220     N = N + 1
230     IF ABS(TERM) > .000001 THEN GOTO 190
240 REM
250     PRINT : PRINT USING "SUM = #.#####"; SUM
260     END
```

Listing 4.4   Program BINOMIAL

Example 1

Use Program BINOMIAL to compute $\sqrt{2}$.

Solution   First we write

$$\sqrt{2} = \sqrt{4 - 2} = 2\sqrt{1 - (1/2)} = 2\left[1 + \left(-\frac{1}{2}\right)\right]^{1/2}.$$

Thus we run Program BINOMIAL with $x = -1/2$ and $k = 1/2$, as shown in Figure 4.5, and then multiply the result by 2.

Note that all variables used are declared as double-precision variables in 140 of Program BINOMIAL. Consequently, we can use the program to compute double-precision roots even if our BASIC (like version 1 of PC-BASIC) has double-precision versions of only the four arithmetical operations, but not of roots and powers.

Programs GEOMETRC and BINOMIAL are based on the same idea:

```
RUN
VARIABLE X? -0.5
EXPONENT K? +0.5
 1
 .75
 .71875
 .7109375
 .70849609375
 .7076416015625
 .7073211669921875
 .7071952819824219
 .7071441411972046
 .7071228325366974
 .7071137763559818
 .707109865732491
 .7071081548347138

SUM = 0.70711

2*SUM = 1.41421
```

Figure 4.5 Computing the square root
of 2 with Program BINOMIAL

In summing the infinite series $\sum c_n$, we calculate the term c_{n+1} by multiplying the preceding term $c_n = c(n)$ by the ratio

$$r(n) = \frac{c(n + 1)}{c(n)}. \tag{13}$$

```
100 REM--Program SERIES
110 REM--Sums a given infinite series after the
120 REM--initial term c(0) is input and the ratio
130 REM--r(n) = c(n+1)/c(n) is edited into line 190.
140 REM
150     PRINT "EDIT IN THE RATIO R(N), THEN RUN 170"
160     PRINT  :  EDIT 190
170     DEFDBL A, N, R, S, T
180     INPUT "INITIAL TERM C(0)"; A
190     DEF FNR(N) =
200     N = 0  :  TERM = A  :  SUM = A
210 REM
220     PRINT N, SUM
230     TERM = TERM*FNR(N)
240     SUM  = SUM + TERM
250     N = N + 1
260     IF ABS(TERM) > .000001 THEN GOTO 220
270 REM
280     PRINT : PRINT USING "SUM = #.#####"; SUM
290     END
```

Listing 4.6 Program SERIES

Program SERIES (Listing 4.6) is a general series summation program that
employs this technique. When the program is run, the ratio $r(n)$ is edited
into line 190 and then used in line 230. Note that the loop for partial sums
(lines 220–260) is otherwise the same as in Program BINOMIAL.

Example 2

To sum the series

$$\sum_{n=0}^{\infty} \frac{1}{(n+1)(n+2)(n+3)} = \frac{1}{1 \cdot 2 \cdot 3} + \frac{1}{2 \cdot 3 \cdot 4} + \frac{1}{3 \cdot 4 \cdot 5} + \cdots,$$

we run Program SERIES with $r(n) = (n+1)/(n+4)$, and input the initial term $c(0) = 1/6 = 0.1666\ldots$. Figure 4.7 shows the result; to avoid printing out 100 partial sums we deleted line 220 and edited GOTO 230 into line 260 before running the program.

```
RUN
EDIT IN THE RATIO R(N), THEN RUN 170

190       DEF FNR(N) = (N + 1)/(N + 4)
RUN 170
INITIAL TERM C(0)? 0.1666666667

SUM = 0.24995
```

Figure 4.7 Summing the series of Example 2

PROBLEMS

Use Program GEOMETRC to sum the geometric series in Problems 1–6, and check your results using Formula (9).

1. $\sum_{n=0}^{\infty} \left(\frac{1}{3}\right)^n$ **2.** $\sum_{n=0}^{\infty} \left(\frac{1}{5}\right)^n$ **3.** $\sum_{n=0}^{\infty} \left(\frac{3}{8}\right)^n$

4. $\sum_{n=0}^{\infty} \left(\frac{7}{17}\right)^n$ **5.** $\sum_{n=0}^{\infty} \left(\frac{9}{10}\right)^n$ **6.** $\sum_{n=0}^{\infty} \left(\frac{59}{69}\right)^n$

Use Program BINOMIAL as in Example 1 to approximate the roots in Problems 7–12. In each case start by finding integers a and h such that

$$\sqrt[n]{A} = \sqrt[n]{a^n + h} = a\left(1 + \frac{h}{a^n}\right)^{1/n}$$

and $|h/a^n| < 1$.

7. $\sqrt{10}$ **8.** $\sqrt{105}$ **9.** $\sqrt[3]{25}$

10. $\sqrt[4]{85}$ **11.** $\sqrt[5]{35}$ **12.** $\sqrt[10]{1000}$

Use Program SERIES to sum the infinite series in Problems 13–20.

13. $e = \sum_{n=0}^{\infty} \frac{1}{n!}$ $(0! = 1)$ **14.** $e^{-1} = \sum_{n=0}^{\infty} \frac{(-1)^n}{n!}$

15. $\sqrt{e} = \sum_{n=0}^{\infty} \frac{1}{2^n \cdot n!}$ **16.** $\ln\frac{3}{2} = \sum_{n=1}^{\infty} \frac{(-1)^{n+1}(1/2)^n}{n}$

17. $\displaystyle\sum_{n=0}^{\infty} n\left(\frac{1}{2}\right)^n$ **18.** $\displaystyle\sum_{n=0}^{\infty} \frac{(-1)^n}{(n+1)(n+2)}$

19. $\displaystyle\sum_{n=0}^{\infty} \frac{n+1}{n!}$ **20.** $\displaystyle\cosh 1 = \sum_{n=0}^{\infty} \frac{1}{(2n)!}$

21. Write a program to input the constants a, b, and c and the variable $|x| < 1$, and then sum the *hypergeometric series*

$$F(a, b, c, x)$$

$$= 1 + \sum_{n=1}^{\infty} \frac{a(a+1)\cdots(a+n-1)b(b+1)\cdots(b+n-1)}{n!\,c(c+1)\cdots(c+n-1)}x^n.$$

Use the fact that $\ln(1+x) = xF(1, 1, 2, -x)$ to test your program.

4.2 EXPONENTIAL AND LOGARITHM FUNCTIONS

Accurate values of the exponential function e^x can be computed using the Taylor series

$$e^x = \sum_{n=0}^{\infty} \frac{x^n}{n!} = 1 + \frac{x}{1!} + \frac{x^2}{2!} + \frac{x^3}{3!} + \cdots. \tag{1}$$

In Section 12.7 of Edwards and Penney, *Calculus* it is shown that for all real x, this series converges to the value e^x. Moreover, the convergence is quite rapid when $|x| \leq 1$. For instance, with $x = 1$ we get the series

$$e = 1 + \frac{1}{1!} + \frac{1}{2!} + \frac{1}{3!} + \cdots. \tag{2}$$

Figure 4.8 shows that when we use Program SERIES to sum this infinite series, we get the value

$$e \approx 2.718281828459045, \tag{3}$$

which is accurate to 15 decimal places, using 18 terms of the series.

If $x > 0$ is large, then series (1) converges less rapidly, so it is better to write

$$e^x = e^{[x]}e^{x-[x]} = e^k e^y, \tag{4}$$

where $k = [x] = \text{Int}(x)$ is the integral part of x (the largest integer not larger than x) and $y = x - k$, so $0 \leq y < 1$. We can then calculate the integral power e^k using the accurate value of e in (3), and compute e^y using series (1). The product of the two results is then the desired value e^x.

This procedure for computing e^x is implemented in Program EXP

```
RUN
EDIT IN THE RATIO R(N), THEN RUN 170

190      DEF FNR(N) = 1/(N + 1)

RUN 170
INITIAL TERM C(0)? 1
 0            1
 1            2
 2            2.5
 3            2.66666666666667
 4            2.70833333333333
 5            2.71666666666667
 6            2.71805555555556
 7            2.718253968253968
 8            2.71827876984127
 9            2.718281525573192
10            2.718281801146385
11            2.718281826198493
12            2.71828182828286169
13            2.718281828446759
14            2.71828182845823
15            2.718281828458995
16            2.718281828459042
17            2.718281828459045
18            2.718281828459045

SUM = 2.718281828459045
```

Figure 4.8 Using Program SERIES to compute the number e to 15 decimal places

(Listing 4.9). Since only the four arithmetical operations $+$, $-$, $\cdot$, and $\div$ are involved, this is in effect a double-precision exponential program for versions of BASIC (such as PC-BASIC 1.0) that do not support double-precision transcendental functions.

The actual Taylor series summation appears in the subroutine in lines 370–430. This subroutine is called first at line 220 to calculate e accurate to 15 decimal places. Then the integral power e^k (with $k = [x]$) is calculated in the FOR-NEXT loop at lines 250–270. The successive integral powers $e^1, e^2, \cdots, e^k$ are stored in the array E(I), so $e^k = $ E(K) when the loop has done its job. The reason we do not simply calculate $e\hat{\ }k$ is that $e\hat{\ }k = \exp(\log(k))$ in BASIC; thus we would be cheating, and moreover we would only get a single-precision result in most versions of BASIC.

Next, the exponential series subroutine is called again at line 300 to compute EY $= e^y$ with $y = x - k$. Finally, the product EX $= e^x = e^y e^k$ is calculated in line 330, and the result printed. All this is done under the assumption that x is positive. The variable F ($= 0$ or 1) is a flag that has been set in line 170 to record whether $x > 0$ or $x < 0$. In case the input x is negative (F $= 1$), we actually compute $\exp(-x)$; then $\exp(x) = 1/\exp(-x)$ is our final result. Figure 4.10 shows the output of some typical runs of Program EXP.

Once we can compute double-precision values of the exponential function $e^x = \exp(x)$, the way is also open to computing double-precision values of the natural logarithm $\ln x = \log(x)$. Given the number $x > 0$, its

```
100 REM--Program EXP
110 REM--Uses the Taylor series to compute the
115 REM--value of exp(x) when x is input.
120 REM
130 REM--Initialization:
140      DEFDBL A,E,S,T,X,Y    :    DEFINT F,I,K,N
150      INPUT "X"; X
160      IF X = 0 THEN EX = 1   :    GOTO 450
170      IF X > 0 THEN F = 0 ELSE F = 1  :   X = -X
180      K = INT(X)   :   DIM E(K)   :   E(0) = 1
190 REM
200 REM--Compute e:
210      IF K = 0 THEN GOTO 300
220      A = 1   :   GOSUB 370   :   E = SUM
230 REM
240 REM--Compute e^k = E(K):
250      FOR I = 1 TO K
260          E(I) = E*E(I-1)
270      NEXT I
280 REM
290 REM--Compute e^y = EY:
300      A = X - K   :   GOSUB 370   :   EY = SUM
310 REM
320 REM--Calculate e^x = EX:
330      EX = EY*E(K)
340      IF F = 1 THEN EX = 1/EX   :   X = -X
350      GOTO 450
360 REM
370 REM--Exp series subroutine:
380      TERM = 1   :   SUM = 1   :   N = 0
390      TERM = TERM*A/(N + 1)
400      SUM  = SUM + TERM
410      N = N + 1
420      IF ABS(TERM) > 1E-16 THEN GOTO 390
430      RETURN
440 REM
450      PRINT "EXP("; X; ") = "; EX
460      END
```

Listing 4.9   Program EXP

```
RUN
X? 1
EXP( 1 ) =  2.718281828459045

RUN
X? 2
EXP( 2 ) =  7.389056098930651

RUN
X? 0.5
EXP( .5 ) =  1.648721270700128

RUN
X? -1
EXP(-1 ) =  .3678794411714423
```

Figure 4.10   Output of Program EXP

natural logarithm is that (unique) number $L = \log(x)$ such that $e^L = x$. Thus we need only use Newton's method to solve the equation

$$f(L) = e^L - x = 0. \qquad (5)$$

Because $f'(L) = e^L$, our Newton iteration is

$$L_{n+1} = L_n - \frac{f(L_n)}{f'(L_n)}$$

$$= L_n - \frac{\exp{(L_n)} - x}{\exp{(L_n)}}$$

$$= L_n - 1 + \frac{x}{\exp{(L_n)}}. \tag{6}$$

To begin the iteration, we need a reasonable initial guess L_0. Supposing that $x > 1$, so that $\log{(x)} > 0$, we compute the integral powers $e^1, e^2, e^3, \ldots$ until we get to the integer m such that $e^m < x \leq e^{m+1}$. Then

$$m < \log{(x)} \leq m + 1, \tag{7}$$

so we use $L_0 = m + 1$ for our initial guess.

```
100 REM--Program LOG
110 REM--Uses Newton's method to solve the equation
120 REM--f(x) = exp(L) - x = 0   for L = log(x).
130 REM
140 REM--Initialization:
150     DEFDBL A,E,L,S,T,X,Y   :   DEFINT F,I,K,M,N
160     INPUT "POSITIVE X"; XO
170     IF XO = 1 THEN LX = 0   :   GOTO 410
180     IF XO > 1 THEN FL = 0 ELSE FL = 1 : XO = 1/XO
190     M = 0   :   EM = 1
200 REM
210 REM--Compute e:
220     A = 1   :   GOSUB 640   :   E = SUM
230 REM
240 REM--Find initial estimate:
250     WHILE  EM < XO
260         EM = E*EM   :   M = M + 1
270     WEND
280     LOLD = M
290 REM
300 REM--Newton's method iteration:
310     PRINT LOLD
320     X = LOLD   :   GOSUB 440
330     EL = EX   :   REM -- EL = e^LOLD
340     LNEW = LOLD - 1 + XO/EL
350     IF ABS(LNEW - LOLD) < 1E-15 THEN GOTO 380
360     LOLD = LNEW   :   GOTO 310
370 REM
380 REM--Print result:
390     LX = LNEW
400     IF FL = 1 THEN LX = -LX   :   XO = 1/XO
410     PRINT "LOG("; XO; ") = "; LX
420     GOTO 720
430 REM
440 REM--EXP Subroutine:
450 REM
460     IF X = O THEN EX = 1   :   GOTO 710
470     IF X > O THEN F = O ELSE F = 1   :   X = -X
```

Listing 4.11 Program LOG

```
480        K =INT(X)    :    DIM E(K)   :    E(0)=1
490 REM
500 REM--Compute e^k = E(K):
510        FOR I = 1 TO K
520            E(I) = E*E(I-1)
530        NEXT I
540 REM
550 REM--Compute e^y = EY:
560        A = X - K   :    GOSUB 640   :    EY = SUM
570 REM
580 REM--Calculate e^x = EX:
590        EX = EY*E(K)
600        ERASE E
610        IF F = 1 THEN EX = 1/EX   :    X = -X
620        GOTO 710
630 REM
640 REM--Exp series subroutine:
650        TERM = 1   :    SUM = 1   :    N = 0
660        TERM = TERM*A/(N + 1)
670        SUM  = SUM + TERM
680        N = N + 1
690        IF ABS(TERM) > 1E-16 THEN GOTO 660
700        REM
710        RETURN
720        END
```

Listing 4.11 (con't.)

This procedure is implemented by Program LOG, shown in Listing 4.11. The input number whose natural logarithm we wish to compute is denoted by X0 because X is reserved for use in the EXP subroutine (lines 440–710). After the number e is computed (line 220), the initial estimate $L_0 = \text{LOLD}$ is computed (as described above) in the WHILE-WEND loop at lines 250–270. Newton's method appears in lines 310–360.

Note that the EXP subroutine is essentially the same as Program EXP in Listing 4.9. Indeed, we constructed Program LOG by first writing the front-end main program (lines 100–420) and then merging EXP. Of course, it was necessary first to save a copy of EXP as EXPA in ASCII format (beginning at line 500), then load the LOG main program, then MERGE "EXPA". After deleting the superfluous lines of EXP, some final editing of the subroutine was required. For instance, because PC-BASIC does not dynamically dimension arrays, the array E(I) of integral powers of e must be redimensioned (line 480) each time the EXP subroutine is called, and then erased (line 600) after it has done its job.

Figure 4.12 shows some typical output of Program LOG.

PROBLEMS

1. Write a front-end main program for Program LOG that converts it into a program—call it EXPLOG—that asks you to input either EXP or LOG, depending on whether you want to compute e^x or $\ln x$, and then proceeds to do whichever you choose.

```
RUN
LOG( 1 ) =  0

RUN
POSITIVE X?  2
 1
 .7357588823428846
 .6940422999189153
 .6931475810597714
 .6931471805600256
 .6931471805599454
LOG( 2 ) =  .6931471805599453

RUN
POSITIVE X?  10
 3
 2.497870683678639
 2.320470381374198
 2.302744085477727
 2.302585105632681
 2.302585092994046
LOG( 10 ) =  2.302585092994046

RUN
POSITIVE X?  2.718281828459045
 1
LOG( 2.718281828459045 ) =  .9999999999999999
```

Figure 4.12   Output of Program LOG

2. The six hyperbolic functions are defined by

$$\cosh x = \frac{e^x + e^{-x}}{2}, \qquad \sinh x = \frac{e^x - e^{-x}}{2},$$

$$\tanh x = \frac{\sinh x}{\cosh x}, \qquad \coth x = \frac{1}{\tanh x},$$

$$\operatorname{sech} x = \frac{1}{\cosh x}, \qquad \operatorname{sech} x = \frac{1}{\sinh x}.$$

Write a program—call it HYPER—that asks you which hyperbolic function of x you wish to compute, and then uses EXP as a subroutine to do so.

3. The six inverse hyperbolic functions are given by

$$\cosh^{-1} x = \ln \left(x + \sqrt{x^2 - 1} \right) \qquad \text{(all } x\text{)},$$

$$\sinh^{-1} x = \ln \left(x + \sqrt{x^2 - 1} \right) \qquad (x \geqq 1),$$

$$\tanh^{-1} x = \frac{\ln (x + 1) - \ln (x - 1)}{2} \qquad (|x| < 1),$$

$$\coth^{-1} x = \frac{\ln (x + 1) - \ln (x - 1)}{2} \qquad (|x| > 1),$$

$$\operatorname{sech}^{-1} x = \ln \left(1 + \sqrt{1 - x^2} \right) - \ln x \qquad (0 < x \leqq 1),$$

$$\operatorname{csch}^{-1} x = \ln \left(|x| + x\sqrt{1 + x^2} \right) - \ln (x|x|) \qquad (x \neq 0).$$

Write a program—call it ARCHYPER—that asks you which of the inverse hyperbolic functions you wish to compute, and then proceeds to do so. The program should include a LOG subroutine to compute natural logarithms, and a Babylonian averaging subroutine to compute double-precision square roots.

4. The logarithmic series

$$\ln(1 + t) = t - \frac{t^2}{2} + \frac{t^3}{3} - \frac{t^4}{4} + \cdots$$

converges if $|t| < 1$, and converges quite rapidly if $|t| < 1/2$. Write a program—call it LNSERIES—that computes $\ln x$ as follows: First find an integer n so that

$$\frac{1}{2} < \frac{x}{2^n} < \frac{3}{2},$$

then use the logarithmic series to compute $\ln(x/2^n)$, and finally add $n \ln 2$ to get $\ln x$. The final step assumes that you have stored an accurate value of $\ln 2$ (found some other way).

5. Write a program to compute $E = e^x$ (when x is input) by applying Newton's method to solve the equation

$$f(E) = \ln E - x = 0.$$

First derive the iterative formula

$$E_{n+1} = E_n(1 - x - \ln E_n).$$

Use Program LNSERIES (Problem 4) as a subroutine to compute the logarithm.

4.3 TRIGONOMETRIC FUNCTIONS

We now discuss the computation of values of trigonometric and inverse trigonometric functions. The Taylor series

$$\cos x = 1 - \frac{x^2}{2!} + \frac{x^4}{4!} - \frac{x^6}{6!} + \cdots \tag{1}$$

and

$$\sin x = x - \frac{x^3}{3!} + \frac{x^5}{5!} - \frac{x^7}{7!} + \cdots \tag{2}$$

are derived in Section 12.7 of Edwards and Penney, *Calculus*. Both series converge for all x (in radians).

However, they converge most rapidly if $|x| \leq 1$. Given x, it is therefore most efficient to find first the smallest nonnegative integer k such that $|x|/2^k \leq 1$, and then use series (1) and (2) to compute the values $\cos(x/2^k)$ and $\sin(x/2^k)$. We can then apply the double-angle formulas

$$\cos 2A = \cos^2 A - \sin^2 A, \tag{3}$$

$$\sin 2A = 2 \sin A \cos A \tag{4}$$

to compute $\cos x$ and $\sin x$. A first application of (3) and (4) yields

$$\cos \frac{x}{2^{k-1}} = \cos^2 \frac{x}{2^k} - \sin^2 \frac{x}{2^k},$$

$$\sin \frac{x}{2^{k-1}} = 2 \sin \frac{x}{2^k} \cos \frac{x}{2^k}.$$

A second application of (3) and (4) yields the values of $\cos(x/2^{k-2})$ and $\sin(x/2^{k-2})$, and after k applications we arrive at the values of $\cos x$ and $\sin x$.

```
100 REM--Program TRIG
110 REM--Uses the sine and cosine Taylor series
115 REM--to compute the sine, cosine, and tangent
120 REM--of the imput value x (radians).
130 REM
140 REM--Initialization:
150     DEFDBL C, S, T, X   :     DEFINT I, K, N
160     INPUT "X"; X
170     K = 0
180     WHILE ABS(X) > 1
190         X = X/2   :   K = K + 1
200     WEND
210 REM
220 REM--Get sine and cosine:
230     GOSUB 400   :   SINE   = SUM
240     GOSUB 500   :   COSINE = SUM
250     IF K = 0 THEN GOTO 600
260     FOR I = 1 TO K
270         S = SINE   :   C = COSINE   :   X = 2*X
280         SINE = 2*S*C   :   COSINE = C*C - S*S
290     NEXT I
300     GOTO 600
310 REM
400 REM--Sine subroutine:
410     TERM = X   :     SUM = X   :   N = 1
420     TERM = -TERM*X*X/((N+1)*(N+2))
430     SUM = SUM + TERM   :   N = N + 2
440     IF ABS(TERM) > 1E-16 THEN   GOTO 420
450     RETURN
```

Listing 4.13 Program TRIG

```
460 REM
500 REM--Cosine subroutine:
510       TERM = 1    :    SUM = 1    :    N = O
520       TERM = -TERM*X*X/((N+1)*(N+2))
530       SUM = SUM + TERM    :    N = N + 2
540       IF ABS(TERM) > 1E-16 GOTO 520
550       RETURN
560 REM
600 REM--Print results:
610       PRINT "SIN("; X; ") = "; SINE
620       PRINT "COS(" X; ") = "; COSINE
630       IF COSINE <> O
             THEN PRINT "TAN("; X ") = "; SINE/COSINE
             ELSE PRINT "TANGENT UNDEFINED"
640       END
```

Listing 4.13 (con't.)

This procedure is implemented in Program TRIG (Listing 4.13). The integer k such that $|x|/2^k \leq 1$ is found in the WHILE-WEND loop at lines 180–200. Then at lines 230 and 240 the sine and cosine subroutines (lines 400–450 and lines 500–550) for the summation of series (1) and (2) are called to compute the sine and cosine of $x/2^k$. The FOR-NEXT loop at lines 260–290 then applies formulas (3) and (4) k times in succession to compute $\sin x$ and $\cos x$. Finally, these values, together with $\tan x = \sin x/\cos x$, are printed at lines 610–630. The results shown in Figure 4.14 indicate that Program TRIG yields 15-decimal-place accuracy.

```
RUN
X?  3.141592653589793
SIN( 3.141592653589793 ) =  2.081668171172169D-16
COS( 3.141592653589793 ) = -1
TAN( 3.141592653589793 ) = -2.081668171172169D-16

PRINT 3.141592653589793/6
 .5235987755982988

RUN
X? -.5235987755982988
SIN(-.5235987755982988 ) = -.5
COS(-.5235987755982988 ) =  .8660254037844387
TAN(-.5235987755982988 ) = -.5773502691896257
```

Figure 4.14   Output of Program TRIG

The Sine Function and the Number π

Perhaps the most elegant approach to trigonometric functions is to *define* $\cos x$ and $\sin x$ by means of the power series (1) and (2). We see immediately that

$$\cos(0) = 1 \quad \text{and} \quad \sin(0) = 0, \tag{5}$$

and termwise differentiation of (1) and (2) yields

$$D \cos x = -\sin x \quad \text{and} \quad D \sin x = \cos x \tag{6}$$

as the derivatives of $\cos x$ and $\sin x$. Then we find that

$$D \left(\cos^2 x + \sin^2 x\right) = 2(\cos x)(-\sin x) + 2(\sin x)(\cos x) = 0,$$

so it follows that

$$\cos^2 x + \sin^2 x = C$$

for some constant C. But (5) gives $C = 1$ when we substitute $x = 0$, so we have derived the fundamental identity of trigonometry,

$$\cos^2 x + \sin^2 x = 1. \tag{7}$$

So far the famous number π, so ubiquitous in trigonometry, has not been mentioned. If we use Program TRIG to compute values of $\sin x$, we find that $\sin x > 0$ for $0 < x \leq 3$, but that $\sin 3.2 < 0$. Hence the interval $[3.0, 3.2]$ contains a solution of the equation

$$f(x) = \sin x = 0. \tag{8}$$

This solution—the smallest positive root of $\sin x$—is, by *definition*, the number π.

```
100 REM--Program PIFIND
110 REM--Computes PI by solving the equation
115 REM--f(x) = sin x = 0 using Newton's method
120 REM--with initial guess X0 = 3.
130 REM
140 REM--Initialization:
150     DEFDBL  X, S, T
160     XOLD = 3 : K = 1
170 REM
180 REM--Newton's iteration:
190 REM
200     X = XOLD  :  GOSUB 290  :   SINE  = SUM
210     X = XOLD  :  GOSUB 300  :  COSINE = SUM
220     XNEW = XOLD - SINE/COSINE
230     PRINT  K, XNEW
240     IF ABS(XNEW - XOLD) < 1E-16 THEN GOTO 370
250     XOLD = XNEW : K = K + 1 : GOTO 200
260 REM
```

Listing 4.15 Program PIFIND

```
270 REM--Sine-cosine subroutine:
280 REM
290         TERM = X   :   SUM = X   :   N = 1   :   GOTO 310
300         TERM = 1   :   SUM = 1   :   N = 0
310         WHILE ABS(TERM) > 1E-16
320             TERM = - TERM*X*X/((N+1)*(N+2))
330             SUM = SUM + TERM   :   N = N + 2
340         WEND
350         RETURN
360 REM
370         END
```

Listing 4.15 (con't.)

Program PIFIND (Listing 4.15) employs Newton's method to find π by solving equation (8) with initial guess $x_0 = 3$. The familiar iteration of Newton's method appears in lines 220–250, with

$$x_{n+1} = x_n - \frac{f(x_n)}{f'(x_n)} = x_n - \frac{\sin x_n}{\cos x_n}.$$

Noting that the sine and cosine subroutines in Program TRIG differ only in their initial lines, we have now combined them in a single subroutine (lines 270–350). Figure 4.16 shows that Program PIFIND yields the value of π accurate to 15 decimal places.

	RUN	
	1	3.14254654267816
	2	3.141592653279601
	3	3.141592653589793
Figure 4.16 Finding π with PIFIND	4	3.141592653589793

Inverse Trigonometric Functions

Here we confine our attention to the most important of the inverse trigonometric functions—the arctangent function arctan x, denoted in BASIC by ATN(X). The other five inverse trigonometric functions can be expressed in terms of the arctangent, and are discussed in the problems.

For each real number x, the inverse tangent of x is the unique number $A = \arctan x$ in the interval $-\pi/2 < A < \pi/2$ whose tangent is x, that is, which satisfies the equation

$$f(A) = \tan A - x = 0. \tag{9}$$

With x fixed, we can solve this equation for A by Newton's method, noting that $f'(A) = \sec^2 A$. Our iterative formula is then

$$A_{n+1} = A_n - \frac{f(A_n)}{f'(A_n)}$$

$$= A_n - \frac{\tan A_n - x}{\sec^2 A_n}$$

$$= A_n - (\cos^2 A_n)(\tan A_n - x)$$

$$A_{n+1} = A_n - \sin A_n \cos A_n + x \cos^2 A_n. \tag{10}$$

```
100 REM--Program ARCTAN
110 REM--Computes y = arctan(x) by using Newton's method
120 REM--to solve the equation f(y) = tan(y) - x = 0.
130 REM
140 REM--Initialization:
150     DEFDBL  A, C, P, X, S, T
160     INPUT "X"; X
170     F = 1  :  IF X < 0 THEN LET F = -1  :  X = -X
180     PI = 3.141592653589793#
190     F1 = 0  :  IF X > 1 THEN F1 = 1  :  X = 1/X
200     AOLD = 1   :   K = 1
210 REM
220 REM--Newton's iteration:
230 REM
240     A = AOLD  :  GOSUB 330  :  SINE  = SUM
250     A = AOLD  :  GOSUB 340  :  COSINE = SUM
260     ANEW = AOLD - SINE*COSINE + X*COSINE*COSINE
270     PRINT  K, ANEW
280     IF ABS(ANEW - AOLD) < 1E-16 THEN GOTO 410
290     AOLD = ANEW : K = K + 1 : GOTO 240
300 REM
310 REM--Sine-cosine subroutine:
320 REM
330     TERM = A  :  SUM = A  :  N = 1  :  GOTO 350
340     TERM = 1  :  SUM = 1  :  N = 0
350     WHILE ABS(TERM) > 1E-16
360         TERM = - TERM*A*A/((N+1)*(N+2))
370         SUM = SUM + TERM  :  N = N + 2
380     WEND
390     RETURN
400 REM
410     ATAN = F*ANEW   :   X = F*X
420     IF F1 = 1 THEN ATAN = PI/2 - ATAN  :  X = 1/X
430     PRINT : PRINT "ARCTAN("; X; ") = "; ATAN
440     END
```

Listing 4.17   Program ARCTAN

This iteration is implemented by Program ARCTAN (Listing 4.17). Since

$$\arctan(-x) = -\arctan x,$$

we can replace x by $-x$ if $x < 0$; the flag F that is set in line 170 tells us whether or not to change back at the end (in line 410). To ensure convergence of Newton's iteration, we also replace x by $1/x$ if $x > 1$, so that we work with $x \leq 1$. The flag F1 that is set in line 190 then tells us whether to apply the identity

$$\arctan x = \operatorname{arccot} \frac{1}{x} = \frac{\pi}{2} - \arctan \frac{1}{x}$$

at line 420. For this purpose the 15-place value of π is stored in line 180. Newton's iteration appears in lines 260–290, and we use the same sine-cosine subroutine as in Program PIFIND. We check this arctangent program by computing $\arctan(1) = \pi/4$ (Figure 4.18). Note that this computation does not use the stored value of π, so we have verified again the 15-place value of π.

```
RUN
X?  1
  1              .837277868313588
  2              .7881802928324653
  3              .7854059179778896
  4              .7853981634575822
  5              .7853981633974483
  6              .7853981633974483

ARCTAN( 1 ) =   .7853981633974483

PRINT 4*ATAN
3.141592653589793
```

Figure 4.18 Using Program ARCTAN to compute the number π

PROBLEMS

1. Revise Program PIFIND so that it computes the number π by using Newton's method to solve the equation $1 + \cos x = 0$.
2. Change Program ARCTAN so as to get a program—call it ARCSIN—that computes $\arcsin x$ rather than $\arctan x$.
3. The other five inverse trigonometric functions are given in terms of $\arctan x$ by

$$\arcsin x = \arctan \frac{x}{\sqrt{1 - x^2}},$$

$$\arccos x = \frac{\pi}{2} - \arcsin x,$$

$$\operatorname{arccot} x = \frac{\pi}{2} - \arctan x,$$

$$\operatorname{arcsec} x = \arccos \frac{1}{x},$$

$$\operatorname{arccsc} x = \arcsin \frac{1}{x}.$$

Write a program—call it INVTRIG—that asks you which of the inverse trigonometric functions you wish to compute, and then proceeds to do so. The program should include an arctan subroutine and a Babylonian averaging subroutine to compute double-precision square roots.

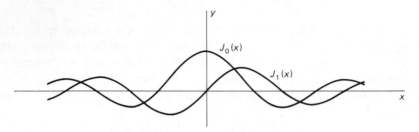

Figure 4.19   Graphs of Bessel functions

4. Figure 4.19 shows the graph $y = J_0(x)$ of the Bessel function of order zero,

$$J_0(x) = \sum_{n=0}^{\infty} \frac{(-1)^n x^{2n}}{2^{2n}(n!)^2}$$

$$= 1 - \frac{x^2}{2^2} + \frac{x^4}{2^2 4^2} - \frac{x^6}{2^2 4^2 6^2} + \cdots .$$

It looks roughly like a damped cosine function, alternating between positive and negative values, but approaching 0 as $x \longrightarrow \infty$. Write a program to tabulate enough values of $J_0(x)$ to verify that its first four positive roots lie in the intervals $[2.4, 2.5]$, $[5.5, 5.6]$, $[8.6, 8.7]$, and $[11.7, 11.8]$.

5. Continuing Problem 4, write a program to compute accurately the first four roots of $J_0(x)$ by using Newton's method to solve the equation $J_0(x) = 0$. The values to four decimal places are 2.4048, 5.5201, 8.6537, and 11.7915. The derivative of $J_0(x)$ is $J_0'(x) = -J_1(x)$, where

$$J_1(x) = \sum_{n=0}^{\infty} \frac{(-1)^n x^{2n+1}}{2^{2n+1} n! \, (n+1)!}$$

$$= \frac{x}{2} - \frac{x^3}{2^3 2!} + \frac{x^5}{2^5 2! \, 3!} - \cdots .$$

4.4 THE ARCHIMEDEAN PROGRAMME

This section is devoted to one last attack on the number π. We shall first retrace, and then extend somewhat, the steps in history's first great computation of π—that of Archimedes in the third century B.C.

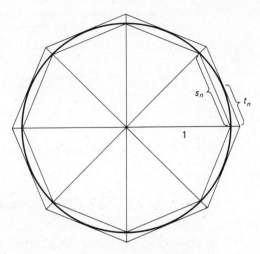

Figure 4.20 Regular *n*-sided polygons inscribed in and circumscribed about the unit circle

Figure 4.20 shows two regular polygons having *n* equal sides, one inscribed in and the other circumscribed about the unit circle (radius = 1). Let P_n and Q_n denote *half* the perimeter of the inscribed polygon and of the circumscribed polygon, respectively. Then, since π is *half* the perimeter of the unit circle, it is apparent (although not quite obvious) that

$$P_n < \pi < Q_n. \tag{1}$$

If somehow we can calculate P_n and Q_n, we therefore will have found lower and upper bounds for the number π.

Archimedes' idea was to start with $n = 6$, in which case the two polygons are regular hexagons. The inscribed one has six sides of length 1, and the circumscribed one has six sides of length $\sec 30° = 2/\sqrt{3}$, so

$$P_6 = 3(1) = 3 \quad \text{and} \quad Q_6 = 3\left(\frac{2}{\sqrt{3}}\right) = 2\sqrt{3}. \tag{2}$$

Then (1) yields the crude estimate $3 < \pi < 2\sqrt{3} \approx 3.464$. Archimedes next doubled the number of sides of the two polygons, and calculated P_{12} and Q_{12}. In fact, he doubled the number of sides four times in succession, calculating first P_{12} and Q_{12}, then P_{24} and Q_{24}, then P_{48} and Q_{48}, and finally P_{96} and Q_{96}. The results he obtained are shown (in decimal form rather than the sexagesimal numeration Archimedes used) in Figure 4.21.

The entries in the P_n column of Figure 4.21 are rounded down to six places, while those in the Q_n column are rounded up, in order for the rounding to respect the direction of the inequalities in (1). From the final

Number n of sides	Semiperimeter Pn of Inscribed Polygon	Semiperimeter Qn of Circumscribed Polygon
6	3.000000	3.464102
12	3.105828	3.215391
24	3.132628	3.159660
48	3.139350	3.146087
96	3.141031	3.142715

Figure 4.21 Archimedes' data

row of data for 96-sided regular polygons we see that $3.1410 < \pi < 3.1428$. Now

$$\frac{223}{71} \approx 3.140845 \quad \text{and} \quad \frac{22}{7} \approx 3.142857,$$

so $P_{96} < \pi < Q_{96}$ yields Archimedes' famous result

$$3\frac{10}{71} < \pi < 3\frac{1}{7}. \tag{3}$$

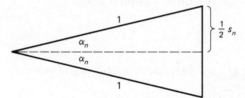

Figure 4.22 A side s_n of the regular n-sided polygon

Now let us explain how Archimedes carried out his computation. We will confine our attention to inscribed polygons, and leave the circumscribed polygons to the problems. Figure 4.22 shows a single side of length s_n of the regular n-sided polygon inscribed in the unit circle. It subtends at the center of the circle an angle $2\alpha_n$, where $\alpha_n = 180°/n$. From the right triangles in the figure we see that

$$s_n = 2 \sin \alpha_n. \tag{4}$$

Once we have found s_n, the semiperimeter of the inscribed polygon is

$$P_n = \frac{1}{2}ns_n = n \sin \alpha_n. \tag{5}$$

Because we know, to begin with, that $s_6 = 1$ is the side of the regular hexagon, the heart of Archimedes' doubling process is the calculation of s_{2n}

in terms of s_n. Since $\alpha_{2n} = \alpha_n/2$, the double-angle sine formula yields

$$
\begin{aligned}
s_n &= 2 \sin \alpha_n = 4 \sin \alpha_{2n} \cos \alpha_{2n} \\
&= 2 \sin \alpha_{2n} \sqrt{4 - 4 \sin^2 \alpha_{2n}} \\
s_n &= s_{2n} \sqrt{4 - s_{2n}^2}
\end{aligned}
$$

because $s_{2n} = 2 \sin \alpha_{2n}$. Squaring this last equation yields the quadratic equation

$$ s_{2n}^4 - 4s_{2n}^2 + s_n^2 = 0 $$

in s_{2n}^2. By the quadratic formula, the root that is small when n is large is

$$ s_{2n}^2 = 2 - \sqrt{4 - s_n^2}. $$

When we multiply and divide the right-hand side by $2 + \sqrt{4 - s_n^2}$, and then take the square root, we get Archimedes' formula,

$$ s_{2n} = \frac{s_n}{\sqrt{2 + \sqrt{4 - s_n^2}}}, \tag{6} $$

for s_{2n} in terms of s_n.

Since $s_1 = 1$, successive applications of (6) yield

$$
\begin{aligned}
s_{12} &= \frac{1}{\sqrt{2 + \sqrt{3}}}, \\
s_{24} &= \frac{1}{\sqrt{(2 + \sqrt{3})(2 + \sqrt{2 + \sqrt{3}})}},
\end{aligned}
$$

and much more complicated formulas for s_{48} and s_{96}. Recalling that $P_{2n} = ns_{2n}$, we can now compute the entries in the second column of Figure 4.21.

Repeated Extrapolation

Archimedes worked without the benefit of algebraic notation, let alone a handy pocket calculator. Impressive though his results were, the data in Figure 4.21 are rather crude by modern standards—they amount to only two-place accuracy, $\pi \approx 3.14$. However, it is an amazing fact that buried somewhere in Archimedes' original data are a full 15 decimal places of π! We can extract them by repeatedly extrapolating in the fashion of the Richardson extrapolation of Section 3.2.

It simplifies the notation to write $A_1, A_2, A_3, \ldots$ instead of P_6, $P_{12}, P_{24}, \ldots$ for our sequence of semiperimeters. Thus A_n is the semiperimeter of the regular polygon with $3 \cdot 2^n$ sides inscribed in the unit circle. Richardson extrapolation yields the sequence $B_2, B_3, B_4, \ldots$ of improved approximations defined by

$$B_n = \frac{1}{3}\left(4A_n - A_{n-1}\right) \tag{7}$$

for $n \geqq 2$.

To see why (7) *is* an improvement, we need to analyze the *error* $\pi - A_n$ made in replacing the exact value π (whatever it is) by the approximation A_n. Formula (5) gives

$$A_n = (3 \cdot 2^n) \sin \frac{\pi}{3 \cdot 2^n}.$$

When we substitute the Taylor series

$$\sin x = x - \frac{x^3}{3!} + \frac{x^5}{5!} - \cdots$$

we get

$$\pi - A_n = \pi - (3 \cdot 2^n)\left[\frac{\pi}{3 \cdot 2^n} - \frac{1}{3!}\left(\frac{\pi}{3 \cdot 2^n}\right)^3 + \frac{1}{5!}\left(\frac{\pi}{3 \cdot 2^n}\right)^5 - \cdots\right]$$

$$= \frac{\pi^3}{3!\, 3^2 2^{2n}} - \frac{\pi^5}{5!\, 3^4 2^{4n}} + \frac{\pi^7}{7!\, 3^6 2^{6n}} \cdots$$

$$\pi - A_n = \frac{a_1}{4^n} + \frac{a_2}{4^{2n}} + \frac{a_3}{4^{3n}} + \cdots, \tag{8}$$

where $a_1, a_2, a_3, \ldots$ are constants whose exact values we do not need (so our discussion does not require that we know π in advance). Thus the error $\pi - A_n$ is of the order of $1/4^n$, corresponding to the initial (largest) term in (8).

Now we analyze the error $\pi - B_n$ in the Richardson extrapolation B_n. From (7) and (8) we get

$$\pi - B_n = \pi - \frac{4}{3}A_n + \frac{1}{3}A_{n-1}$$

$$= \frac{4}{3}(\pi - A_n) - \frac{1}{3}(\pi - A_{n-1})$$

$$= \frac{4}{3}\left(\frac{a_1}{4^n} + \frac{a_2}{4^{2n}} + \frac{a_3}{4^{3n}} + \cdots\right)$$

$$- \frac{1}{3}\left(\frac{a_1}{4^{n-1}} + \frac{a_2}{4^{2n-2}} + \frac{a_3}{4^{3n-3}} + \cdots\right)$$

$$\pi - B_n = \frac{b_2}{4^{2n}} + \frac{b_3}{4^{3n}} + \frac{b_4}{4^{4n}} + \cdots, \tag{9}$$

where $b_2, b_3, b_4, \ldots$ are constants. Comparing (8) and (9), we see that the error $\pi - B_n$ is of the order $1/4^{2n}$, a significant improvement over $\pi - A_n$ of order $1/4^n$.

This process of extrapolation can be continued. If we define the second extrapolations $C_3, C_4, C_5, \ldots$ by

$$C_n = \frac{1}{15}(16B_n - B_{n-1}) \tag{10}$$

for $n \geq 3$, a calculation similar to the one above yields

$$\pi - C_n = \frac{c_3}{4^{3n}} + \frac{c_4}{4^{4n}} + \frac{c_5}{4^{5n}} + \cdots, \tag{11}$$

where $c_3, c_4, c_5, \ldots$ are constants. Thus the error $\pi - C_n$ is of the order of $1/4^{3n}$, another significant improvement.

Two more extrapolations are worth our while. The third extrapolations $D_4, D_5, D_6, \ldots$ are defined by

$$D_n = \frac{1}{63}(64C_n - C_{n-1}) \tag{12}$$

for $n \geq 4$, and it turns out that

$$\pi - D_n = \frac{d_4}{4^{4n}} + \frac{d_5}{4^{5n}} + \cdots. \tag{13}$$

The fourth extrapolations $E_5, E_6, E_7, \ldots$ are defined by

$$E_n = \frac{1}{255}(256D_n - D_{n-1}) \tag{14}$$

for $n \geq 5$, and it transpires that

$$\pi - E_n = \frac{e_5}{4^{5n}} + \frac{e_6}{4^{6n}} + \cdots. \tag{15}$$

Whatever the constant e_5 may be, the sequence $\{e_5/4^{5n}\}$ will approach 0 very rapidly as $n \longrightarrow \infty$, so the fourth extrapolations $\{E_n\}$ should be extremely good approximations to π.

```
100 REM--Program ARCHIMED
110 REM--Approximates pi by repeated extrapolation
115 REM--from the semiperimeters of regular polygons
120 REM--inscribed in the unit circle,
130 REM--starting with the regular hexagon.
140 REM
150 REM--Initialization:
160     DEFDBL A, B, C, D, E, S, X, Y
170     DIM  A(10), B(10), C(10), D(10), E(10)
180     S = 1  :  N = 6  :  K = 5  :  A(1) = 3
190 REM
200 REM--S equals a side and, for I<=K, A(I) is
205 REM--the semiperimeter of an inscribed regular
210 REM--polygon with 3 times 2^I sides.
220 REM
230 REM--Loop to compute semiperimeters:
240     FOR I=1 TO K
250         PRINT "A(";I;") = "; A(I)
260         A = 4 - S*S : GOSUB 630
270         A = 2 + B   : GOSUB 630
280         S = S/B
290         A(I+1) = N*S
300         N = 2*N
310     NEXT I
320     PRINT
330 REM
340 REM--First extrapolation:
350     FOR I=2 TO K
360         B(I) = (4*A(I) - A(I-1))/3
370         PRINT "B(";I;") = "; B(I)
380     NEXT I
390     PRINT
400 REM
410 REM--Second extrapolation:
420     FOR I=3 TO K
430         C(I) = (16*B(I)-B(I-1))/15
440         PRINT "C(";I;") = "; C(I)
450     NEXT I
460     PRINT
470 REM
480 REM--Third extrapolation:
490     FOR I=4 TO K
500         D(I) = (64*C(I)-C(I-1))/63
510         PRINT  "D(";I;") = "; D(I)
520     NEXT I
530     PRINT
540 REM
550 REM--Fourth extrapolation:
560     FOR I=5 TO K
570         E(I) = (256*D(I)-D(I-1))/255
580         PRINT "E(";I;") = "; E(I)
590     NEXT I
600 REM
610     GOTO 720
620 REM
```

Listing 4.23   Program ARCHIMED

```
630 REM--Babylonian square root subroutine
635 REM--which returns B = SQR(A):
640 REM
650      X = A
660      Y = (X + A/X)/2
670      IF ABS (X-Y) < 1E-15 THEN GOTO 690
680      X = Y : GOTO 660
690      B = Y
700      RETURN
710 REM
720      END
```

Listing 4.23 (con't.)

Program ARCHIMED (Listing 4.23) carries out this process of repeated extrapolation for Archimedes' original inscribed regular polygons with 6, 12, 24, 48, and 96 sides. The double-precision arrays A(I), B(I), C(I), D(I), and E(I) are established in line 170 to hold the semiperimeters $\{A_n\}$ and the successive extrapolations. The original semiperimeters are calculated in the FOR-NEXT loop at lines 240–310; note that lines 260–280 implement the basic doubling formula in (6). The following four loops carry out the four successive extrapolations. Note that lines 360, 430, 500, and 570 correspond to Formulas (7), (10), (12), and (14) defining the extrapolations. Double-precision square roots are computed by the Babylonian subroutine in lines 650–700.

Figure 4.24 shows the amazing output of Program ARCHIMED. Our single fourth extrapolation E(5) gives correctly the first 15 decimal places of π, all of them extracted from just the five polygons that Archimedes used. To get this accuracy without extrapolation, it would be necessary to double the number of sides 25 times in succession, winding up with a regular polygon having $3 \cdot 2^{25} = 100,663,296$ sides!

```
RUN
A( 1 ) = 3
A( 2 ) = 3.105828541230249
A( 3 ) = 3.132628613281238
A( 4 ) = 3.139350203046867
A( 5 ) = 3.14103195089051

B( 2 ) = 3.141104721640332
B( 3 ) = 3.141561970631568
B( 4 ) = 3.141590732968744
B( 5 ) = 3.141592533505057

C( 3 ) = 3.14159245389765
C( 4 ) = 3.141592650457889
C( 5 ) = 3.141592653540811

D( 4 ) = 3.141592653577892
D( 5 ) = 3.141592653589747

E( 5 ) = 3.141592653589793
```

Figure 4.24 Results of repeated extrapolation of the Archimedean data

PROBLEMS

1. If t_n denotes the length of a side of a regular n-sided polygon circumscribed about the unit circle, show that

$$t_n = 2 \tan \alpha_n,$$

where $\alpha_n = 180°/n$. Then apply the double-angle tangent formula $\tan 2A = (2 \tan A)/(1 - \tan^2 A)$ to derive the doubling formula

$$t_{2n} = \frac{2t_n}{2 + \sqrt{4 + t_n^2}}.$$

2. Use the formula of Problem 1 to amend Program ARCHIMED so that it works with circumscribed rather than inscribed polygons.

3. Deduce from the formulas

$$P_n = n \sin \alpha_n \quad \text{and} \quad Q_n = n \tan \alpha_n$$

for the inscribed and circumscribed semiperimeters the recursive formulas

$$Q_{2n} = \frac{2P_nQ_n}{P_n + Q_n} \quad \text{and} \quad P_{2n} = P_nQ_{2n}$$

for P_{2n} and Q_{2n} in terms of P_n and Q_n.

4. Write a program that employs the recursive formulas of Problem 3 to produce the data shown in Figure 4.21.

5. Use the program of Problem 4 to estimate π by starting with inscribed and circumscribed squares, and doubling the number of sides 10 times in succession to obtain regular polygons with 4096 sides.

6. Verify Formulas (11), (13), and (15) in the text.

Velocity, Acceleration, and Differential Equations

5

5.1 BALLS AND SKYDIVERS

Suppose that a ball is thrown straight upward with initial velocity v_0 at time $t = 0$. Then, ignoring air resistance, its velocity $v(t)$ at time t satisfies the differential equation

$$\frac{dv}{dt} = -g, \tag{1}$$

where $g = 32$ ft/sec^2 (approximately) is the acceleration of gravity near the surface of the earth. Its height $y(t)$ at time t satisfies the differential equation

$$\frac{dy}{dt} = v. \tag{2}$$

These equations are easily integrated (as in Section 4.9 of Edwards and Penney, *Calculus*) to obtain the equations

$$v = -gt + v_0, \tag{3}$$

$$y = -\frac{1}{2}gt^2 + v_0 t + y_0 \tag{4}$$

for the ball's velocity and height (with y_0 denoting its initial height).

For example, suppose that $y_0 = 0$ and $v_0 = 160$ ft/sec. Then the ball reaches its maximum height $y_{\max}$ when

$$v = -32t + 160 = 0,$$

that is, after $t = 5$ sec. Then Equation (4) yields

$$y_{\max} = -16(5)^2 + (160)(5) = 400 \text{ ft.}$$

The ball hits the ground when

$$y = -16t^2 + 160t = -16t(t - 10) = 0,$$

and thus at time $t = 10$ sec. The velocity with which it hits the ground is

$$v = -32(10) + (160) = -160 \text{ ft/sec.}$$

In short, the ball ascends in 5 sec to its maximum height of 400 ft, and 5 sec later hits the ground with a speed of 160 ft/sec (the same speed with which it left the ground).

Now we want to explore a more realistic mathematical model that takes air resistance into account. If we assume that the force of air resistance on the ball is proportional to its velocity v, then the acceleration equation in (1) is replaced with

$$\frac{dv}{dt} = -g - kv, \tag{5}$$

where k is an empirical constant. In Section 7.6 of Edwards and Penney, *Calculus*, Equation (5) is solved for the velocity function $v(t)$, which then is integrated to obtain the height function $y(t)$.

Here, however, our aim is to describe a simple procedure by which the ball's motion—under the combined influence of gravity and air resistance—can be approximated numerically with any desired degree of accuracy. We want to approximate the ball's position and velocity—starting with $y = y_0$ and $v = v_0$ at time $t_0 = 0$—at times $t_1, t_2, t_3, \ldots$ with $t_{i+1} = t_i + h$. Thus we will update our approximations every h seconds. The smaller we choose h (and hence the more work we do) the more accurate our approximations ought to be.

Let y_i and v_i denote the ball's approximate height and velocity at time t_i. Equation (5) gives its acceleration then as

$$a_i = -g - kv_i. \tag{6}$$

If the ball experienced a *constant* acceleration of a_i ft/sec^2 during the h seconds from time t_i to time t_{i+1}, its velocity would be changed by $a_i h$ feet per second. We assume that h is so small that $a \approx a_i$ throughout the time interval $[t_i, t_{i+1}]$, and therefore take

$$v_{i+1} = v_i + a_i h \tag{7}$$

as our estimate of the ball's velocity at time t_{i+1}. Its (approximate) *average* velocity during $[t_i, t_{i+1}]$ is then

$$\frac{1}{2}(v_i + v_{i+1}) = v_i + \frac{1}{2}a_i h,$$

so we use

$$y_{i+1} = y_i + \frac{1}{2}(v_i + v_{i+1})h$$

$$= y_i + v_i h + \frac{1}{2}a_i h^2 \tag{8}$$

as our estimate of its height at time t_{i+1}.

```
100 REM--Program BALL
110 REM--Computes height and velocity of vertically
120 REM--thrown ball under the influence of gravity
130 REM--and air resistance proportional to velocity.
140 REM
150 REM--Initialization:
160      N = 0  :  T = 0   :  Y = 0  :  V = 160
170      G = 32 :  H = .1  :  K = 0
180      PRINT "    T            V            Y" : PRINT
190 REM
200 REM--Beginning of iteration:
210      WHILE Y > -50
220         IF N/10 = N\10 THEN PRINT USING
            "  ###       +###.##        ###.##"; T,V,Y
230         A  = -G - K*V
240         Y  = Y + V*H + .5*A*H*H
250         V  = V + A*H
260         T  = T + H  :  N = N + 1
270      WEND
280 REM
290      END
```

Listing 5.1   Program BALL

Program BALL (Listing 5.1) computes the approximations described above. Lines 230, 240, and 250 correspond to Equations (6), (8), and (7) respectively. We take $h = 0.1$ to update every tenth of a second, but (because of the condition tested in line 220) print only every tenth line of data. We do not know in advance how long the ball will be in the air, so we continue the iteration as long as $y > -50$ (line 210), so as to follow the ball until just after it returns to ground level ($y = 0$).

RUN

T	V	Y
0	+160.00	0.00
1	+128.00	144.00
2	+96.00	256.00
3	+64.00	336.00
4	+32.00	384.00
5	+0.00	400.00
6	-32.00	384.00
7	-64.00	336.00
8	-96.00	256.00
9	-128.00	144.00
10	-160.00	0.00

Figure 5.2 Run of Program BALL with $k = 0$

We initially set $k = 0$ in line 170 because we want to test the program in the case of no air resistance. The results shown in Figure 5.2 agree with our earlier exact computation—the ball is in the air for 10 sec and reaches a maximum height of 400 ft.

Example 1

If the ball is thrown straight upward from ground level with initial velocity $v_0 = 160$ ft/sec, and is subject to air resistance with $k = 0.1$ in Equation (5), what is its maximum height, and how long is it in the air?

RUN

T	V	Y
0	+160.00	0.00
1	+114.10	136.67
2	+72.60	229.68
3	+35.06	283.19
4	+1.11	300.99
5	-29.60	286.49
6	-57.36	242.78
7	-82.48	172.65
8	-105.19	78.63
9	-125.73	-37.00

Figure 5.3 Run of Program BALL with $k = 0.1$

Solution Figure 5.3 shows the results of a run of Program BALL. It indicates a maximum height of just over 300 ft and impact with the ground at some time between $t = 8$ and $t = 9$ sec. To focus in on the maximum height and impact, we reran Program BALL with the print control line changed to

220 IF (N > 39 AND N < 46) OR (N > 84 AND N < 91)
    THEN PRINT USING "   #.#   +###.##   ###.##"; T,V,Y

The results shown in Figure 5.4 indicate that the ball reaches a maximum height of just over 301 ft in just over 4.0 sec, and that it hits the ground after about 8.7 sec with a speed of about 120 ft/sec. These results agree (to within 1 ft and 1/10 sec) with an exact solution of the problem. Note that the effect of air resistance is to reduce the ball's maximum height, its impact velocity, and the duration of its flight (as compared with no air resistance).

RUN

T	V	Y
4.0	+1.11	300.99
4.1	-2.10	300.94
4.2	-5.28	300.57
4.3	-8.43	299.89
4.4	-11.55	298.89
4.5	-14.63	297.58
8.5	-115.72	23.38
8.6	-117.76	11.71
8.7	-119.78	-0.17
8.8	-121.78	-12.25
8.9	-123.77	-24.53
9.0	-125.73	-37.00

Figure 5.4 More detailed data with $k = 0.1$

In the case of a large body moving fairly rapidly through the air, it is generally more realistic to assume that the air resistance is proportional to the *square* of the velocity. In this case Equation (5) is replaced with

$$a = \frac{dv}{dt} = -g - kv|v|, \tag{9}$$

where k is an empirical constant. We write $v|v|$ instead of v^2 so that the equation will apply whether the ball is moving upward (resistance force downward) or downward (resistance force upward). In Problem 1 we ask you to investigate the motion of the thrown ball of Example 1 if the air resistance is proportional to v^2.

Example 2

Investigate the descent of a skydiver who bails out of an airplane at 10,000 ft. Assume that $k = 0.0032$ in Equation (9); this value is reasonable if the skydiver assumes the "spread eagle" position and wears loose-fitting clothing to increase the air resistance.

Solution Here it is convenient to take the y-axis pointing downward, with $y = 0$ at the plane and $y = 10,000$ at the ground. Then the gravitational acceleration is positive while the air resistance acceleration is negative, so we use

$$a = \frac{dv}{dt} = g - kv^2 \tag{10}$$

instead of Equation (9). Listing 5.5 shows Program SKYDIVER, which is essentially the same as Program BALL except that line 230 now corresponds to Equation (10). We print data every 2 sec for the first 20 sec; the results are shown in Figure 5.6.

The most interesting feature of the descent is that the skydiver rapidly reaches a "terminal velocity" of 100 ft/sec. At this point the gravitational acceleration and air resistance cancel, and no further change in velocity occurs. Note that after only 6 sec the skydiver has reached 96 percent of terminal velocity.

```
100 REM--Program SKYDIVER
110 REM--Computes downward velocity and position of
120 REM--skydiver with air resistance proportional to
130 REM--the square of the velocity.
140 REM
150 REM--Initialization:
160     N = 0   :   T = 0   :   Y = 0   :   V = 0
170     G = 32  :   H = .1  :   K = .0032
180     PRINT  "     T              V              Y": PRINT
190 REM
200 REM--Beginning of iteration:
210     WHILE  T < 21
220         IF  N/20 = N\20 THEN PRINT USING
            "   ###        ###.##       ####.##";  T,V,Y
230         A  =  G - K*V*V
240         Y  =  Y + V*H  + .5*A*H*H
250         V  =  V + A*H
260         T  =  T + H   :   N = N + 1
270     WEND
280 REM
290     END
```

Listing 5.5   Program SKYDIVER

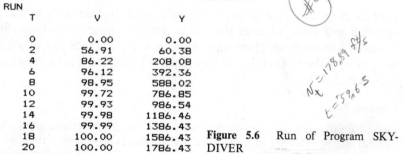

```
RUN
     T          V               Y

     0        0.00            0.00
     2       56.91           60.38
     4       86.22          208.08
     6       96.12          392.36
     8       98.95          588.02
    10       99.72          786.85
    12       99.93          986.54
    14       99.98         1186.46
    16       99.99         1386.43
    18      100.00         1586.43
    20      100.00         1786.43
```

Figure 5.6   Run of Program SKY-
DIVER

After 20 sec the skydiver has fallen 1786 ft, and falls the remaining 8214 ft (without benefit of parachute) to the ground in about 82 sec, hitting the ground 1 min 42 sec after bailout. With an impact velocity of 100 ft/sec ≈ 68 mph (miles per hour) and with the aid of a strategically located haystack, he or she might even survive!

PROBLEMS

Use appropriate alterations of Programs BALL and SKYDIVER to solve the following problems numerically.

1. Suppose that the ball of Example 1 experiences air resistance proportional to v^2 with $k = 0.001$ in Equation (9). Then find its maximum height, duration of flight, and impact velocity. The values of all three should be comparable to those found in Example 1 with air resistance proportional to v.

2. If the ball of Problem 1 (v^2 resistance with $k = 0.001$) is dropped from an airplane at a height of 10,000 ft, find its terminal velocity and the time it takes to reach the ground.

3. The same as Problem 2, except that the ball is thrown straight downward with an initial velocity of 150 ft/sec.

4. A crossbow bolt is shot straight upward with an initial velocity of 288 ft/sec. Find its maximum height if (a) there is no air resistance; (b) there is air resistance proportional to v with $k = 0.04$.

5. The same as Problem 4(b), except that the air resistance is proportional to v^2 with $k = 0.0002$.

6. A motorboat weighs 32,000 lb, and its motor provides a thrust of 5000 lb. Assume that the water resistance is 100 pounds for each foot per second of the speed v of the boat. Then

$$100 \frac{dv}{dt} = 5000 - 100v.$$

If the boat starts from rest, what is the maximum speed it can attain?

7. A man bails out of an airplane at an altitude of 10,000 ft, falls freely for 20 sec, then opens his parachute. Assume air resistance proportional to v with $k = 0.15$ without parachute and $k = 1.5$ with parachute. How long will it take him to reach the ground? *Suggestion:* First find his altitude and velocity when the parachute first opens.

The following three problems are adapted from Section 5.5 of W. R. Bennett, Jr., *Scientific and Engineering Problem-Solving with the Computer* (Englewood Cliffs, N.J.: Prentice-Hall, Inc., 1976).

8. The acceleration a of a 4.2-liter E-type Jaguar is given by

$$a = \frac{dv}{dt} = 5.6 \exp\left(\frac{125 - v}{145}\right) - 0.22 - 5.6\left(\frac{v}{125}\right)^2,$$

where v is in miles per hour and t is in seconds. How long does it take to accelerate from rest to 120 mph?

9. Suppose that a U-2 pilot bails out at an altitude of 76,000 ft and that his downward acceleration is

$$a = \frac{dv}{dt} = g - k\, d(y)v^2,$$

where $y = 32.16$ ft/sec^2, $k = 0.0006$, and the factor

$$d(y) = \begin{cases} \exp\left(-3.21 \times 10^{-5}y\right) & \text{if } y \le 30{,}000 \text{ ft}, \\ 1.60 \exp\left(-4.77 \times 10^{-5}y\right) & \text{if } y > 30{,}000 \text{ ft} \end{cases}$$

represents the diminution of air density with altitude. How long will it take the pilot to free-fall to an altitude of 10,000 ft?

10. The 1500-hp engine of a 123-ton diesel locomotive generates an accelerating force of

$$F(v) = \begin{cases} 62{,}500 \text{ lb} & \text{if } 0 \le v \le 8 \text{ mph,} \\ \dfrac{500{,}000}{v} \text{ lb} & \text{if } v > 8 \text{ mph.} \end{cases}$$

Its acceleration dv/dt (in feet per second per second) satisfies the equation

$$m\frac{dv}{dt} = F(v) - R(v),$$

where $m = 7640$ slugs and the resistive force is

$$R(v) = 277 + 3.72v + 0.24v^2$$

pounds (v in miles per hour). What is the maximum velocity $v_{\max}$ that this locomotive can attain? How long will it take to reach a speed of $0.95v_{\max}$?

5.2 PROJECTILE TRAJECTORIES

The modern age of computing dates back (at least) to the construction of the ENIAC (Electronic Numerical Integrator and Computer) during World War II. It contained 18,000 vacuum tubes plus 80,000 capacitors and resistors, and measured 100 ft by 10 ft by 3 ft, but had nothing like the power and speed of the IBM-PC. The ENIAC was designed specifically to compute projectile trajectories like those discussed in this section, although our examples are phrased in terms of baseballs rather than artillery shells.

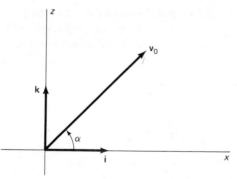

Figure 5.7   The initial velocity vector $\mathbf{v}_0$ with inclination angle α

We consider a projectile that is fired from the origin in the vertical xz-plane (Figure 5.7) with its initial velocity vector $\mathbf{v}_0$ inclined at an angle α to the horizontal x-axis. Thus its initial velocity components are

$$v_{0x} = v_0 \cos \alpha \quad \text{and} \quad v_{0z} = v_0 \sin \alpha, \tag{1}$$

where $v_0 = |\mathbf{v}_0|$ is its initial speed. We suppose that the projectile moves under the influence of gravity and of air resistance, with the acceleration $\mathbf{a}_R$ of resistance being directed opposite to the velocity vector $\mathbf{v} = v_x \mathbf{i} + v_z \mathbf{k}$ and having magnitude proportional to the square of the speed $v = |\mathbf{v}|$. Then, since $\mathbf{v}/|\mathbf{v}|$ is a unit vector in the direction of $\mathbf{v}$,

$$\mathbf{a}_R = -kv^2 \frac{\mathbf{v}}{|\mathbf{v}|} = -kv\mathbf{v}$$

$$= -kvv_x \mathbf{i} - kvv_z \mathbf{k}, \tag{2}$$

where k is an empirical constant (not to be confused with the unit vector $\mathbf{k}$ in the z-direction). Consequently, the projectile's acceleration vector is

$$\mathbf{a} = a_x \mathbf{i} + a_z \mathbf{k} = \mathbf{a}_R - g\mathbf{k},$$

$$= (-kvv_x)\mathbf{i} + (-kvv_z - g)\mathbf{k}, \tag{3}$$

where $g \approx 32$ ft/sec^2 denotes gravitational acceleration. Consequently, the motion of the projectile is described by the differential equations

$$\frac{d^2x}{dt^2} = -kvv_x \quad \text{and} \quad \frac{d^2z}{dt^2} = -kvv_z - g, \tag{4}$$

where $v_x = dx/dt$, $v_z = dz/dt$, and $v = (v_x^2 + v_z^2)^{1/2}$.

If air resistance is ignored, so that $k = 0$, then two integrations of the equations in (4) yield

$$x = v_{0x}t + x_0,$$

$$z = -\frac{1}{2}gt^2 + v_{0z}t + z_0 \tag{5}$$

for the position (x, z) of the projectile at time t. For instance, suppose the projectile is a baseball that leaves the bat at the origin with initial speed 160 ft/sec and initial inclination angle 30°. Then $v_{0x} = 160 \cos 30° = 80\sqrt{3}$ and $v_{0z} = 160 \sin 30° = 80$, so equations (5) yield

$$x = 80\sqrt{3}\, t \quad \text{and} \quad z = -16t^2 + 80t.$$

Hence the baseball returns to the ground $(z = 0)$ after $t = 5$ sec, having traveled a horizontal distance of $x = 80\sqrt{3}\,(5) = 400\sqrt{3} \approx 692.82$ ft.

When air resistance is included, so that $k > 0$, the differential equations in (4) are quite difficult to solve exactly, and we resort to the numerical approximation technique of Section 5.1, applying it separately to the horizontal x-components and the vertical z-components of motion. That is, with (s_i, v_i, a_i) denoting either the triple (x, v_x, a_x) of x-components or the triple (z, v_z, a_z) of z-components at time t_i, we use the equations

$$v_{i+1} = v_i + a_i h \qquad\qquad (6)$$

and

$$s_{i+1} = s_i + v_i h + \frac{1}{2} a_i h^2 \qquad\qquad (7)$$

[corresponding to Equations (7) and (8) in Section 5.1] to update the values to time $t_{i+1} = t_i + h$.

```
100 REM--Program BASEBALL
110 REM--Computes trajectory of baseball with air
120 REM--resistance proportional to velocity squared.
130 REM
140 REM--Initialization:
150      X = 0 : Z = 0 : T = 0 : N = 0
160      INPUT "INITIAL VELOCITY"; V
170      INPUT "INITIAL INCLINATION (DEG)"; A
180      PI = 3.141593   :   A = PI*A/180
190      VX = V*COS(A)   :   VZ = V*SIN(A)
200      K = 0           :   G = 32    :    H = .01
210      PRINT " T          X         Z          V         A"
220      PRINT
230 REM
240 REM--Beginning of iteration:
250      WHILE Z > -5
260          V = SQR (VX*VX + VZ*VZ)
270          IF  N/50 <> N\50  THEN GOTO 290
280          PRINT USING
                "#.#      ####.##       ###.##     ###.##     +##";
                T, X, Z, V, 180*ATN(VZ/VX)/PI
290          AX = - K*V*VX   :     AZ = - K*V*VZ - G
300          X  =  X + VX*H  +  .5*AX*H*H
310          Z  =  Z + VZ*H  +  .5*AZ*H*H
320          VX = VX + AX*H  :     VZ = VZ + AZ*H
330          T  =  T + H     :     N  = N + 1
340      WEND
350 REM
360      END
```

Listing 5.8   Program BASEBALL

Program BASEBALL (Listing 5.8) carries out these computations. Lines 160 and 170 call for us to input the initial velocity v_0 and the initial angle of inclination α. Line 290 calculates the acceleration components a_x and a_z corresponding to the differential equations in (4). Lines 300–320 correspond to Equations (6) and (7). The data are updated each $h = 0.01$ sec, but because of the condition in line 270, we print data only every 1/2 sec (when the counter N is a multiple of 50). Because we do not know in advance how long the ball will remain aloft, line 260 directs that the iterations continue while $z > -5$, and hence stop only after the ball has hit the ground ($z = 0$).

```
RUN
INITIAL VELOCITY? 160
INITIAL INCLINATION (DEG)? 30
   T          X            Z           V           A

  0.0        0.00        0.00       160.00       +30
  0.5       69.28       36.00       152.63       +25
  1.0      138.56       64.00       146.64       +19
  1.5      207.85       84.00       142.21       +13
  2.0      277.13       96.00       139.48        +7
  2.5      346.41      100.00       138.56        +0
  3.0      415.69       96.00       139.48        -7
  3.5      484.98       84.00       142.21       -13
  4.0      554.26       64.00       146.64       -19
  4.5      623.54       36.00       152.63       -25
  5.0      692.82       -0.00       160.00       -30
```

Figure 5.9   Run of Program BASEBALL with $k = 0$

Figure 5.9 shows the results of a run of Program BASEBALL with $k = 0$ (no air resistance), initial velocity $v_0 = 160$ ft/sec, and initial inclination angle $\alpha = 30°$. The last two columns of the table show the current velocity and the angle of inclination of the velocity vector (a negative angle meaning that the ball is descending). We see that the ball reaches a maximum height of 100 ft, remains aloft for 5 sec, and travels a horizontal distance of 692.82 ft before hitting the ground. Thus the results shown in Figure 5.9 are in full agreement with the exact equations of motion in (5).

```
RUN
INITIAL VELOCITY? 160
INITIAL INCLINATION (DEG)? 30
   T          X            Z           V           A

  0.0        0.00        0.00       160.00       +30
  0.5       63.22       32.72       127.08       +24
  1.0      117.05       53.14       104.74       +17
  1.5      164.20       63.49        89.61        +7
  2.0      206.31       65.14        80.07        -3
  2.5      244.38       59.00        75.15       -15
  3.0      279.02       45.78        73.96       -27
  3.5      310.59       26.08        75.47       -37
  4.0      339.31        0.51        78.69       -46
```

Figure 5.10   Run of Program BASEBALL with $k = 0.0025$

Figure 5.10 shows the results of a run of Program BASEBALL with the realistic air resistance coefficient $k = 0.0025$. The last line of data printed is just before impact ($z \approx 1/2$ ft). Now the ball reaches a maximum height of only about 66 ft, is aloft just over 4 sec, and travels only about 340 ft. Thus air resistance has converted a massive home run into a routine fly ball (if hit straightaway to centerfield). Note also that the ball returns to the ground with only about half its initial velocity, and that its descent is considerably steeper than its ascent. The massive home run and the routine flyball are compared in Figure 5.11.

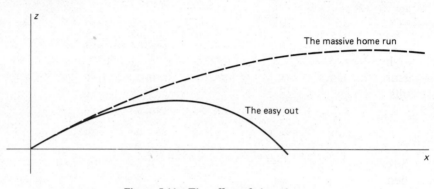

Figure 5.11   The effect of air resistance

Does a Baseball Pitch Really Curve?

How often have you wondered whether a baseball curve is "real" or whether it is some sort of optical illusion? To investigate this question, let's suppose that the pitcher fires the ball in the x-direction toward home plate (which is 60 ft away; Figure 5.12), and gives it a spin of S revolutions per second, counterclockwise (as viewed from above) about a vertical axis through the ball's center. This spin is described by the *spin vector* **S** shown in Figure 5.13; **S** points along the axis of revolution in the right-handed direction, and has length $S = |\mathbf{S}|$.

Figure 5.12 x-axis pointing toward home plate

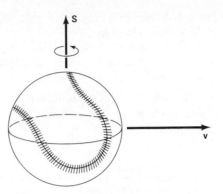

Figure 5.13 The baseball's spin and
velocity vectors

In aerodynamics it is shown that this spin causes a difference in the air
pressures on the sides of the ball toward and away from the spin, and that this
results in a *spin acceleration*

$$\mathbf{a}_S = c\mathbf{S} \times \mathbf{v} \quad \text{(cross product)} \tag{8}$$

of the ball (where c is an empirical constant). The total acceleration of the ball
is then

$$\mathbf{a} = \mathbf{a}_G + \mathbf{a}_R + \mathbf{a}_S, \tag{9}$$

where $\mathbf{a}_G = -g\mathbf{k}$ is the gravitational acceleration and

$$\mathbf{a}_R = -kv(v_x\mathbf{i} + v_y\mathbf{j} + v_z\mathbf{k}) \tag{10}$$

is the acceleration of air resistance proportional to v^2.

The cross-product vector in (8) is given (see Section 14.2 of Edwards
and Penney, *Calculus*) by

$$\mathbf{S} \times \mathbf{v} = \begin{vmatrix} \mathbf{i} & \mathbf{j} & \mathbf{k} \\ S_x & S_y & S_z \\ v_x & v_y & v_z \end{vmatrix}$$

$$= (S_y v_z - S_z v_y)\mathbf{i} + (S_z v_x - S_x v_z)\mathbf{j} + (S_x v_y - S_y v_x)\mathbf{k}, \tag{11}$$

where $S = S_x\mathbf{i} + S_y\mathbf{j} + S_y\mathbf{k}$. For instance, if $\mathbf{S} = S\mathbf{k}$ as in the case of the
pitched baseball described above, then $S_x = S_y = 0$ and $S_z = S$, so (11)
yields

$$\mathbf{S} \times \mathbf{v} = -Sv_y\mathbf{i} + Sv_x\mathbf{j}. \tag{12}$$

For a ball pitched along the x-axis, v_x is much larger than v_y. Indeed, if $v_y = 0$, then (8) reduces to

$$\mathbf{a}_S = cSv_x\mathbf{j}, \tag{13}$$

so the spin acceleration is directed to the left (looking in toward home plate). In essence, this accounts for the fact that a baseball pitch actually can curve.

```
100 REM--Program CURVBALL
110 REM--Computes trajectory of baseball with both spin
120 REM--acceleration and air resistance proportional
130 REM--to velocity squared.
140 REM
150 REM--Initialization:
160     DIM   X(3), V(3), A(3), G(3), S(3), SXV(3)
170     INPUT "INITIAL POSITION X,Y,Z";    X(1),X(2),X(3)
180     INPUT "INITIAL VELOCITY VX,VY,VZ"; V(1),V(2),V(3)
190     INPUT "SPIN VECTOR SX,SY,SZ";      S(1),S(2),S(3)
200     T  = 0      :   H = .005
210     K  = .0025  :   C = .005   :   G(3) = -32
220     PRINT " T          X           Y           Z          V"
230     PRINT
240 REM
250 REM--Beginning of iteration:
260     FOR N = 0 TO 110
270        V = SQR (V(1)*V(1) + V(2)*V(2) + V(3)*V(3))
280        IF   N < 100   AND   N/10 <> N\10   THEN GOTO 300
285        IF   N/2 <> N\2   THEN GOTO 300
290        PRINT USING
           "#.##      ###.##     ###.##     ###.##     ###";
           T, X(1), X(2), X(3), V
300        SXV(1) = S(2)*V(3)  -  S(3)*V(2)  :
           SXV(2) = S(3)*V(1)  -  S(1)*V(3)  :
           SXV(3) = S(1)*V(2)  -  S(2)*V(1)
310        FOR   I = 1 TO 3
320           A(I) = C*SXV(I)  -  K*V*V(I) + G(I)
330           X(I) = X(I) + V(I)*H  +  .5*A(I)*H*H
340           V(I) = V(I) + A(I)*H
350        NEXT I
360        T  =  T + H
370     NEXT N
380 REM
390     END
```

Listing 5.14   Program CURVBALL

Program CURVBALL (Listing 5.14) computes the trajectory of a ball under the influence of gravity, spin acceleration, and air resistance proportional to v^2. So as not to require separate program lines for the calculation of the individual components of the position, velocity, and acceleration vectors, we use $x_1x_2x_3$-coordinates instead of xyz-coordinates. The principal horizontal direction of motion is denoted by x_1, the perpendicular direction in the horizontal plane by x_2, and the vertical direction by x_3. Three-element arrays to hold the components of the position, velocity, acceleration, gravity

$(g_1 = g_2 = 0, g_3 = 32)$, spin, and $\mathbf{S} \times \mathbf{v}$ vectors are dimensioned in line 160. Lines 170–190 prompt us to input the initial position, the initial velocity vector, and the spin vector. Line 300 calculates the components of the cross-product vector $\mathbf{S} \times \mathbf{v}$ as defined in Equation (11). The loop in lines 310–350 then updates the acceleration, position, and velocity vectors—lines 320, 330, and 340 correspond to Equations (9), (7), and (6), respectively.

```
RUN
INITIAL POSITION X,Y,Z? 0,0,5
INITIAL VELOCITY VX,VY,VZ? 120,-3,4
SPIN VECTOR SX,SY,SZ? 0,0,40
   T            X            Y            Z           V

  0.00         0.00         0.00         5.00        120
  0.05         5.96        -0.12         5.16        118
  0.10        11.83        -0.18         5.24        117
  0.15        17.61        -0.18         5.23        115
  0.20        23.31        -0.12         5.15        113
  0.25        28.93        -0.01         4.99        112
  0.30        34.47         0.16         4.75        110
  0.35        39.94         0.38         4.43        109
  0.40        45.32         0.65         4.04        107
  0.45        50.64         0.97         3.58        106
  0.50        55.88         1.34         3.04        105
  0.51        56.91         1.42         2.92        105
  0.52        57.95         1.50         2.81        104
  0.53        58.98         1.58         2.68        104
  0.54        60.01         1.66         2.56        104
  0.55        61.04         1.75         2.43        104
```

Figure 5.15   A real curveball

Figure 5.15 shows the results of a run of Program CURVBALL to emulate a carefully planned baseball pitch. The pitcher lets fly with the ball from the pitching mound $(x_1 = x_2 = 0)$ with initial height $x_3 = 5$ ft, with a spin of $S = 40$ rev/sec, and with initial velocity vector $v_0 = \langle 120, -3, 4 \rangle$, so $v_0 \approx 120$ ft/sec ≈ 82 mph. We use the values $k = 0.0025$ for the resistance coefficient and $c = 0.005$ for the spin coefficient [in Equation (8)]. Although the data are updated every $h = 0.005$ sec (line 200), the print control conditions in lines 280–285 direct that data be printed only every $1/20$ sec during the ball's first $1/2$ sec of flight, and then every $1/100$ sec during its final approach to home plate.

To see how this pitch looks to the batter, let's assume that he gets a "fix" on the ball and prepares to swing by observing it during the first $1/4$ sec. During this $1/4$ sec the ball continues to appear as though it is headed straight for home plate at a height of 5 ft. But the final lines in Figure 5.15 show that during the last $1/4$ sec, the ball suddenly tails off over 2 ft downward and 20 in. to the left (away from a right-handed batter). Although the values $c = 0.005$ and $S = 40$ may slightly exceed those attainable by a typical pitcher with an "unloaded" ball, it is now clear that a baseball curve really curves, and is no optical illusion!

PROBLEMS

Problems 1–3 deal with a batted baseball having initial velocity 160 ft/sec. Use Program BASEBALL with air resistance coefficient $k = 0.0025$. You may want to add a print control condition that causes data to be printed more frequently as the ball nears the ground.

1. Find the *range*—the horizontal distance the ball travels before it hits the ground—with initial inclination angles 40°, 45°, and 50°.

2. Find (to the nearest degree) the initial inclination angle that maximizes the range. If there were no air resistance it would be 45°, but your answer should be less than 40°.

3. Find (to the nearest half-degree) the initial inclination angle greater than 45° for which the range is 300 ft.

4. Find the initial velocity of a baseball hit by Babe Ruth (with $k = 0.0025$ and initial inclination 40°) if it hit the bleachers at a point 50 ft high and 500 horizontal feet from home plate.

5. Run Progam CURVBALL with $k = 0$ and $c = 0$ so as to duplicate the data in Figure 5.9.

6. Run Program CURVBALL with $k = 0.0025$ and $c = 0$ so as to duplicate the data in Figure 5.10.

7. Consider a batted baseball with initial velocity 160 ft/sec and initial inclination angle 30°. Run Program CURVBALL with $k = 0.0025$ and $c = 0.005$ to show that a topspin of 40 rev/sec decreases its range by about 80 ft, whereas a backspin of 40 rev/sec increases its range by about 80 ft. Thus a backspin could convert an easy out into a home run.

8. Show that a sidespin of 40 rev/sec—like the spin of the pitched baseball of this section—will cause the batted ball of Problem 7 to drift over 100 ft off a straight (horizontal) line path. In particular, if it is drifting toward the foul line, it must have been batted about 20° inside the foul line in order to stay in fair territory.

9. In the presence of a wind with velocity vector **w**, the air resistance acceleration of a projectile is

$$\mathbf{a}_R = -k\,|\mathbf{v} - \mathbf{w}|\,(\mathbf{v} - \mathbf{w}) \tag{14}$$

instead of $\mathbf{a}_R = -k\,|\mathbf{v}|\,\mathbf{v}$. That is, the absolute velocity vector **v** is replaced with the velocity vector $\mathbf{v} - \mathbf{w}$ relative to the wind. Now suppose that the batted ball of Problem 7 is hit (without spin) along the x-axis, but that a 15-mph (22-ft/sec) crosswind is blowing in the direction of the y-axis. Alter Program CURVBALL in accordance with Equation (14) and run it with $k = 0.0025$, $c = 0$ to determine where the ball hits the ground.

10. The Big Bertha of World War I fame had a 112-ft barrel and a 230-lb shell. The Germans fired it at Paris from a distance of 72.25 miles. Assuming an initial inclination angle of 40°, what initial velocity was required in order for the shell to reach Paris? Ignoring spin and diminution of gravitational acceleration with altitude, use $g = 32.18$ ft/sec^2, $k = 1.2 \times 10^{-5}$, and the air density function of Problem 9 in Section 5.1.

5.3 MISSILES AND ROCKETS

In this section we investigate the motion of missiles and rockets that—in contrast to the projectiles of Section 5.2—venture sufficiently far from the surface of the earth that the inverse-square variation of gravitational attraction must be taken into account. Let M and R denote the mass and radius, respectively, of the earth (assuming it to be spherical). Then the gravitational acceleration of a body at distance r from the center of the earth is directed toward the earth's center and has magnitude

$$a_G = \frac{GM}{r^2}, \tag{1}$$

where G is the universal gravitational constant. Since $a_G = g \approx 32.2 \text{ ft/sec}^2$ at the surface of the earth, where $r = R \approx 3960$ miles, we see that $GM = gR^2$. If y denotes the body's *altitude*—its height above the surface of the earth, substitution of $GM = gR^2$ and $r = R + y$ in (1) yields

$$a_G = \frac{gR^2}{(R + y)^2} = \frac{g}{(1 + y/R)^2}. \tag{2}$$

This is the form in which we will find it most convenient to apply Newton's law of gravitation.

One of history's earliest moonshot plans was described by Jules Verne in 1865. A huge artillery gun was to fire a missile containing three astronauts and two dogs straight toward the moon from a point in Florida near Cape Canaveral(!). To compute the trajectory of such a missile, we need to take into account the variations both of gravity and of air resistance with altitude. We may approximate the variation of air density with altitude by

$$D(y) = \begin{cases} \exp(-0.169y) & \text{if } y < 6, \\ 1.6 \exp(-0.252y) & \text{if } 6 \leqq y \leqq 40, \\ 0 & \text{if } y > 40. \end{cases} \tag{3}$$

Here $D(y)$ is the air density at an altitude of y miles, relative to unit density at sea level $(y = 0)$. The acceleration due to air resistance at altitude y and velocity v is then

$$a_R = kD(y)v^2. \tag{4}$$

Program MOONSHOT (Listing 5.16) enables us to check the feasibility of a scheme like Verne's. After we input the initial velocity (in miles per second) of a vertically projected missile, its altitude and velocity are updated every $h = 0.1$ sec, although data are printed only every 10 sec (line 250).

```
100 REM--Program MOONSHOT
110 REM--Computes the trajectory of a vertically
120 REM--launched projectile under the influence of
130 REM--inverse square gravitation and variable air
140 REM--resistance.  Units are in miles and seconds.
150 REM
160 REM--Initialization:
170        INPUT "INITIAL VELOCITY (MI/SEC)"; V
180        Y =  0  :   T = 0   :   N = 0
190        H = .1  :   V = 1   :   R = 3960
200        G = 32.2/5280       :   K = .000012*5280
210        PRINT "    T           V           Y"   :
           PRINT
220 REM
230 REM--Beginning of iteration:
240        WHILE  Y > -2
250           IF N/100 = N\100 THEN PRINT USING
              "  ###       +#.###       ###.##"; T,V,Y
260           AG = -G/(1 + Y/R)^2
270           IF Y < 6 THEN D = EXP(-.169*Y)
                      ELSE IF  Y > 40   THEN  D = 0
                      ELSE D = 1.6*EXP(-.252*Y)
280           AR = -K*D*V*ABS(V)
290           A  =  AG + AR
300           Y  =  Y  + V*H + .5*A*H*H
310           V  =  V  + A*H
320           T  =  T  + H   :   N  =  N + 1
330        WEND
340 REM
350        END
```

Listing 5.16 Program MOONSHOT

Note that this program is very similar to our original Program BALL in Section 5.1 (Listing 5.1), except that the inverse-square law of gravitation appears in line 260; the air density approximation of (3) appears in line 270, and v^2 air resistance in line 280. We use the air resistance coefficient $k = 0.000012$ (in feet-pound-second units) that corresponds to an aerodynamically designed artillery shell, and convert to miles per second units by means of the 5280-ft/mile factors in line 200.

Figure 5.17 shows the results of a shot with initial velocity 1 mile/sec—roughly the muzzle velocity of the WWI Big Bertha—with time t in seconds, velocity v in miles per second, and altitude y in miles. We observe that the projectile attains its maximum altitude shortly after time $t = 120$ sec and returns to the ground shortly before time $t = 250$ sec. Figure 5.18 shows the results of a second run in which data were printed every second (that is, when $N/10 = N \backslash 10$) when either $120 \leq t \leq 125$ or $245 \leq t \leq 250$. We now see that the projectile reaches a maximum altitude of 45.3 miles in about 121 sec, and then takes about 125 sec to fall back to the ground. In particular, it falls considerably short of its intended lunar destination. In Section 4.9 of Edwards and Penney, *Calculus* it is shown that the "escape velocity" from the surface of the earth is approximately 7 miles/sec.

The data in Figure 5.18 reveal an interesting consequence of the

T	V	Y
0	+1.000	0.00
10	+0.702	8.10
20	+0.613	14.62
30	+0.548	20.42
40	+0.486	25.59
50	+0.426	30.15
60	+0.366	34.11
70	+0.306	37.47
80	+0.246	40.23
90	+0.186	42.39
100	+0.127	43.96
110	+0.067	44.93
120	+0.008	45.30
130	−0.052	45.08
140	−0.112	44.26
150	−0.171	42.84
160	−0.231	40.83
170	−0.291	38.22
180	−0.351	35.01
190	−0.411	31.20
200	−0.471	26.80
210	−0.530	21.79
220	−0.588	16.20
230	−0.633	10.07
240	−0.618	3.73
250	−0.500	−1.94

Figure 5.17 Moonshot with initial velocity 1 mi/sec

T	V	Y
---	------	------
---	------	------
---	------	------
120	+0.008	45.30
121	+0.002	45.30
122	−0.004	45.30
123	−0.010	45.29
124	−0.016	45.28
125	−0.022	45.26
---	------	------
---	------	------
---	------	------
245	−0.571	0.75
246	−0.558	0.18
247	−0.545	−0.37
248	−0.531	−0.91
249	−0.516	−1.43
250	−0.500	−1.94

Figure 5.18 The apex and conclusion of the attempted moonshot

decrease with altitude of the earth's atmospheric density. If the atmospheric density were constant, then—like the skydiver of Section 5.1—the projectile on its return trip would fall with steadily increasing speed until it effectively reached its limiting velocity. However, it actually gains speed more rapidly as it falls through the rarer upper atmosphere, reaches a maximum speed at an altitude of about 10 miles, and then is decelerated as it falls the remaining distance to the ground through air of increasing density.

Rocket Flight

One of the engineers who worked on the German V-2 rocket during World War II is quoted as saying that "We aimed for the moon but hit London by mistake." Here we investigate the motion of such a rocket that is launched vertically (like a moonshot).

In contrast with a projectile that instantaneously acquires an initial velocity $v_0 > 0$ at time $t = 0$, a rocket lifts off with initial velocity $v_0 = 0$ and is propelled upward by the thrust force that is generated by the rearward expulsion of exhaust gases from its burning fuel. If fuel is consumed at a constant rate b (in mass/time units) during the time interval $0 \leq t \leq t_1$, then the variable mass $m(t)$ of the rocket at time t is given by

$$m(t) = \begin{cases} m_0 - bt & \text{if } t < t_1, \\ m_0 - bt_1 & \text{if } t \geq t_1, \end{cases} \tag{5}$$

where m_0 is the rocket's initial mass (including fuel). We assume that the resulting thrust force F_0 is constant while $t < t_1$, and is zero thereafter.

Both before and after burnout this rocket is subject to a force F_R of air resistance and to a downward gravitational force

$$F_G = m(t)a_G, \tag{6}$$

where a_G is the gravitational acceleration at altitude y as given by Equation (2). Then Newton's second law of motion $(F = ma)$ yields

$$m(t) \frac{dy}{dt} = F_0 - F_R - m(t)a_G \tag{7}$$

while $t < t_1$.

A realistic description of the resistance force F_R is rather complex. According to Bennett (see the reference cited in the Section 5.1 Problems),

$$F_R = kD(y)C_D Sv|v| \qquad \text{pounds} \tag{8}$$

where $k = 0.00119$ in feet-pound-second units, $D(y)$ is the relative density of the atmosphere at altitude y feet, C_D is the rocket's drag coefficient, S its (transverse) cross-sectional area, and v its velocity in feet per second. Upon converting y to feet in (3), we get (approximately)

$$D(y) = \begin{cases} \exp\left(-3.21 \times 10^{-5}y\right) & \text{if } y < 30,000, \\ 0 & \text{if } y > 200,000, \\ 1.6 \exp\left(-4.77 \times 10^{-5}y\right) & \text{otherwise.} \end{cases} \tag{9}$$

The drag coefficient C_D depends on the Mach number $M_a = v/1043$—the rocket's speed as a multiple of the speed of sound (1043 ft/sec) at standard conditions. Bennett gives

$$C_D = \begin{cases} 0.1 & \text{if } M_a < 0.89, \\ 1.231M_a - 1 & \text{if } 0.89 < M_a < 1.13, \\ 0.141 + 0.129(M_a^2 - 1)^{-1/2} & \text{if } M_a > 1.13 \end{cases} \qquad (10)$$

for a V-2 rocket. The major effect of (10) is that the drag coefficient C_D increases sharply as the rocket accelerates through the "sound barrier" near Mach 1, but then decreases if the rocket continues to accelerate thereafter.

Finally, the V-2 rocket had an initial weight of 27,000 lb, including 19,200 lb of fuel, which burned for 64 sec at the rate of 300 lb/sec, generating a thrust of 55,000 lb. It was 47 ft long and had a cross-sectional radius of 2.46 ft.

```
100 REM--Program MOONROCK
110 REM--Computes the trajectory of a vertically launched
120 REM--V-2 rocket under the influence of inverse square
130 REM--gravitation and variable air resistance.  Uses
140 REM--ft-lb-sec units in the computation, but converts
150 REM--to miles and seconds for program output.
160 REM
170 REM--Initialization:
180      WO   =   27000      'Initial weight of rocket
190      G    =   32.2       'Surface gravitational acc
200      M    =   WO/G       'Initial mass of rocket (slugs)
210      B    =   300/G      'Burn rate (slugs/sec)
220      FO   =   55000!     'Constant thrust (lbs)
230      TO   =   64         'Burn time (sec)
240      S    =   19.01      'Cross-sectional area (sq ft)
250      K    =   .00119     'Air resistance coefficient
260      R    =   3960*5280  'Radius of earth (ft)
270      Y = 0  :   T = 0   :   N = 0   :   H = .1
280      PRINT "    T           V            Y"   :   PRINT
290 REM
300 REM--Beginning of iteration:
310      WHILE  Y > -3000
320         IF Y < 5000 AND V < 0 THEN P = 10
                ELSE P = 200
330         IF  N/P = N\P  THEN PRINT USING
                "    ###      +#.##        ###.##"; T,V/5280,Y/5280
340         MACH  = ABS(V)/1043
350         IF MACH < .89 THEN C = .1 ELSE
                IF MACH < 1.13 THEN C = 1.231*MACH - 1
                   ELSE C = .141 + .129*(MACH*MACH - 1)^(-.5)
360         IF Y < 30000 THEN D = EXP(-.0000321*Y) ELSE
                IF  Y > 200000! THEN  D = 0
                   ELSE D = 1.6*EXP(-.0000477*Y)
370         AG  =  G/(1 + Y/R)^2
380         FR  =  K*C*D*S*V*ABS(V)
390         IF T < TO THEN FT = FO ELSE FT = 0
400         A   =  - AG + (FT - FR)/M
```

Listing 5.19 Program MOONROCK

```
410          Y  =  Y  + V*H  +  .5*A*H*H
420          V  =  V  + A*H
430          T  =  T  + H    :    N  =  N + 1
440          IF T  <  TO  THEN M = M – B*H
450      WEND
460 REM
470      END
```

Listing 5.19 (con't.)

Program MOONROCK (Listing 5.19) was written to determine how high a vertically launched V-2 rocket will go (if, indeed, it fails to reach the moon). The needed numerical data (in feet-pound-second units) are listed with descriptive comments in lines 180–260. Lines 350 and 360 calculate the drag coefficient C and the atmospheric density D using the formulas in (10) and (9), respectively. Lines 380 and 390 provide the resistance force FR [Equation (8)] and the thrust force FT, and then line 400 calculates the acceleration $a = dv/dt$ from Equation (7). Lines 410–430 update the position Y, velocity V, and time T as usual, and line 440 accounts for the decrease of mass M while the rocket fuel is burning. Lines 320 and 330 control the printing of data—every 20 sec until the final mile of descent, and then every second.

T	V	Y
0	+0.00	0.00
20	+0.16	1.46
40	+0.37	6.61
60	+0.79	17.67
80	+0.83	35.07
100	+0.71	50.40
120	+0.59	63.34
140	+0.47	73.92
160	+0.35	82.15
180	+0.24	88.04
200	+0.12	91.60
220	+0.00	92.82
240	−0.11	91.71
260	−0.23	88.28
280	−0.35	82.51
300	−0.46	74.39
320	−0.58	63.93
340	−0.70	51.11
360	−0.82	35.90
380	−0.94	18.32
400	−0.71	0.27
401	−0.67	−0.42

Figure 5.20   The vertically launched V-2 rocket

Figure 5.20 shows the results of a run of Program MOONROCK. We see immediately that our program name was overly optimistic: The rocket climbs only to an altitude of about 93 miles in 3 min 40 sec, and then takes about 3 min to fall back to earth. Incidentally, Program MOONROCK runs a bit slower than "real time," requiring about 12 sec of computation for each

10 sec of rocket flight. Thus a BASIC compiler would make repeated execution of this program (as in Problem 5.8) more convenient.

Flat Earth Rocket Trajectories

Finally, we consider two-dimensional motion of a rocket near the surface of the earth. However, we ignore the curvature of the earth's surface, so our trajectories will enjoy reasonable accuracy only for rather short flights—when the horizontal range is less than a couple of hundred miles, perhaps.

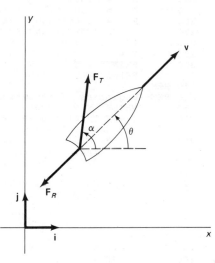

Figure 5.21   The thrust angle α and the direction angle θ

We assume for simplicity that the rocket is controlled in such a way that the thrust force $\mathbf{F}_T$ makes a *constant* angle α with the horizontal (as indicated in Figure 5.21), while the rocket itself remains aligned in the direction of its velocity vector $\mathbf{v}$. Then Equation (7) is replaced with the vector equation

$$m(t)\mathbf{a} = \mathbf{F}_T - \mathbf{F}_R - m(t)\mathbf{a}_G, \tag{11}$$

where $\mathbf{a}_G = g\mathbf{j}$ and

$$\mathbf{F}_T = (F_0 \cos \alpha)\mathbf{i} + (F_0 \sin \alpha)\mathbf{j}, \tag{12}$$

$$\mathbf{F}_R = (F_R \cos \theta)\mathbf{i} + (F_R \sin \theta)\mathbf{j}. \tag{13}$$

Program FLATROCK (Listing 5.22) is a two-dimensional version of Program MOONROCK. We now must calculate separately the two scalar components of the position, velocity, acceleration, and force vectors. The thrust angle $\alpha = $ ANG is input at line 270, and the direction angle

```
100 REM--Program FLATROCK
110 REM--Computes the trajectory of a V-2 rocket with
120 REM--constant thrust angle, under the influence of
130 REM--variable gravitation and air resistance.  Uses
140 REM--ft-lb-sec units in the computation, but converts
150 REM--to miles and seconds for program output.
160 REM
170 REM--Initialization:
180      WO  =   27000        'Initial weight of rocket
190      G   =   32.2         'Surface gravitational acc
200      M   =   WO/G         'Initial mass of rocket (slugs)
210      B   =   300/G        'Burn rate (slugs/sec)
220      FO  =   55000!       'Constant thrust (lbs)
230      TO  =   64           'Burn time (sec)
240      S   =   19.01        'Cross-sectional area (sq ft)
250      K   =   .00119       'Air resistance coefficient
260      R   =   3960*5280    'Radius of earth (ft)
270      INPUT "THRUST ANGLE (DEG)"; ANG
280      PI = 3.141593     :     ANG  =   PI*ANG/180
290      Y = 0   :   T = 0   :   N = 0   :   H = .1
300      PRINT"   T         V          X         Y       ANGLE":
         PRINT
310 REM
320 REM--Beginning of iteration:
330      WHILE  Y > -3000
340        IF THETA < 0 AND Y < 5000 THEN P = 10
             ELSE P = 200
350        IF   N/P = N\P  THEN   PRINT USING
           "  ###    #.##      ###.##    ###.##     +##";
           T, V/5280, X/5280, Y/5280, THETA*180/PI
360        MACH  = ABS(V)/1043
370        IF MACH < .89 THEN C = .1 ELSE
             IF MACH < 1.13 THEN C = 1.231*MACH - 1
               ELSE C = .141 + .129*(MACH*MACH - 1)^(-.5)
380        IF Y < 30000 THEN D = EXP(-.0000321*Y) ELSE
             IF  Y > 200000! THEN  D = 0
               ELSE D = 1.6*EXP(-.0000477*Y)
390        AG =   G/(1 + Y/R)^2
400        FR =   K*C*D*S*V*V
410        IF T < TO THEN FT = FO ELSE FT = 0
420        AX = (FT*COS(ANG) - FR*COS(THETA))/M
430        AY = (FT*SIN(ANG) - FR*SIN(THETA))/M   - AG
440        X  = X  + VX*H  + .5*AX*H*H
450        Y  = Y  + VY*H  + .5*AY*H*H
460        VX = VX  + AX*H
470        VY = VY  + AY*H   :   THETA = ATN(VY/VX)
480        V  =   SQR(VX*VX + VY*VY)
490        T  =   T  + H   :   N  =  N + 1
500        IF T  <  TO  THEN M = M - B*H
510      WEND
520 REM
530      END
```

Listing 5.22   Program FLATROCK

θ = THETA is calculated in line 470. In addition to the speed v, the horizontal distance x, and the altitude y we print—at 20-sec intervals until near impact—the angle θ, so a glance at the data suffices to tell when the rocket is ascending ($\theta > 0$) and when it is descending ($\theta < 0$).

Figure 5.23 shows the data resulting from a run of Program FLATROCK with thrust angle $\alpha = 57°$. We see that the rocket reaches a

T	V	X	Y	ANGLE
0	0.00	0.00	0.00	+0
20	0.19	1.45	1.03	+37
40	0.41	5.97	4.69	+41
60	0.80	14.35	12.62	+45
80	0.85	26.89	24.99	+42
100	0.77	39.50	35.12	+35
120	0.71	52.10	42.85	+27
140	0.66	64.70	48.20	+18
160	0.64	77.31	51.16	+8
180	0.63	89.91	51.74	−3
200	0.65	102.51	49.95	−13
220	0.68	115.11	45.78	−23
240	0.74	127.72	39.22	−32
260	0.81	140.32	30.28	−39
280	0.88	152.89	18.95	−45
300	0.83	164.91	5.81	−50
309	0.63	169.05	0.62	−53
310	0.60	169.42	0.13	−53
311	0.57	169.77	−0.34	−54

Figure 5.23   The V-2 rocket with thrust angle 57°

maximum altitude of about 52 miles and that its horizontal range is about 169.5 miles, with impact occurring after 5 min 10 sec of flight. The program now requires about 28 sec of run time for every 10 sec of flight time, so not only a BASIC compiler but also an 8087 numeric coprocessor might be useful for repeated execution.

PROBLEMS

1. If both air resistance and the variation of gravity with altitude were ignored, a projectile launched vertically with initial velocity 1 mile/sec would reach a maximum altitude of about 82 miles in about 164 sec. Check these results by running Program MOONSHOT with $K = 0$ in line 200 and $AG = -G$ in line 260.

2. If both air resistance and the variation of gravity with altitude are ignored, the maximum altitude attained by a vertically launched projectile with initial velocity v_0 is $y_{max} = v_0^2/2g$. Hence doubling v_0 quadruples y_{max}. Compare this with the situation when air resistance and gravitational variation are included, by running Program MOONSHOT with $v_0 = 2$ miles/sec.

3. If air resistance is ignored but the variation of gravity with altitude is included, a projectile launched vertically with initial velocity 1 mile/sec should (according to theory) reach a maximum altitude of 83.72 miles. Check this result by running Program MOONSHOT with $K = 0$ in line 200.

4. If air resistance is ignored, the escape velocity from the earth's surface is about 6.95 miles/sec. That is, a projectile launched vertically with this initial velocity will continue indefinitely to recede from the earth, instead of attaining a maximum altitude and then falling back to the surface. Now we want to

find the escape velocity when the resistance of the earth's atmosphere is taken into account. Use Program MOONSHOT to determine what the projectile's initial velocity should be, in order that it will have a velocity of 6.95 miles/sec when it reaches an altitude of 40 miles (where the atmosphere has "thinned out" and is no longer a factor).

Use Program MOONROCK (Listing 5.19) with appropriate alterations to solve Problems 5–8.

5. Find the maximum speed attained by the vertically launched V-2 rocket (a) during its ascent; (b) during its descent.

6. Find the vertically launched V-2 rocket's impact velocity (accurate to within 0.01 mile/sec).

7. Suppose that a new fuel doubles (to 110,000 lb) the thrust force of the V-2 rocket. Now what is its maximum altitude if launched vertically, with everything else the same?

8. Assuming that the fuel still burns for 64 sec at the rate of 300 lb/sec, what constant thrust force must it generate in order for the vertically launched V-2 to attain an escape velocity of 6.95 miles/sec at an altitude of 40 miles (see Problem 4)?

9. Write a program to compute the trajectory of a vertically launched *two-stage rocket* with initial mass m_0. During the ith stage $(i = 1, 2)$ fuel burns at rate b_i for t_i seconds, generating a force of F_i pounds. At the end of stage i the remaining mass is m_i. Add any other embellishments you wish—such as differing radii for the two stages and the payload.

Use Program FLATROCK (Listing 5.22) with appropriate alterations to solve Problems 10–12.

10. With thrust angle $\alpha = 57°$ as in Figure 5.23, find the maximum speed attained by the V-2 (a) during its ascent; (b) during its descent.

11. Find the horizontal range of the V-2 with thrust angle 57° if its fuel efficiency is doubled as in Problem 7.

12. Run Program FLATROCK with several different thrust angles so as to convince yourself that the horizontal range is maximal when $\alpha = 57°$ (to the nearest degree).

5.4 PLAIN ORDINARY DIFFERENTIAL EQUATIONS

The first three sections of this chapter have been devoted to the numerical solution of differential equations of motion. We now discuss the general first-order initial value problem

$$\frac{dy}{dx} = f(x, y), \qquad y(x_0) = y_0 \tag{1}$$

independently of its physical origin or intended application (if any).

If the function $f(x, y)$ has continuous partial derivatives $\partial f / \partial x$ and $\partial f / \partial y$ in a neighborhood of the point (x_0, y_0), then it is known that on some open interval about the point x_0 on the x-axis there exists a unique solution $y(x)$ of the differential equation $y' = f(x, y)$ that satisfies the initial condition $y(x_0) = y_0$. However, unless the differential equation happens to be either linear or separable (see Sections 18.2 and 18.3 of Edwards and Penney, *Calculus*), it is likely that no explicit formula for the solution $y(x)$ can be found. In this event we must resort to numerical approximation techniques.

The simplest such technique is *Euler's method* for computing numerical approximations to the solution of (1). We first choose a fixed *step size* $h > 0$ and consider the points

$$x_0, x_1, x_2, \ldots, x_n, \ldots, \tag{2}$$

where $x_n = x_0 + nh$, so $x_n = x_{n-1} + h$, for each $n = 1, 2, 3, \ldots$. Our goal is to find suitable approximations $y_1, y_2, y_3, \ldots$ to the true values of the solution $y(x)$ at the points $x_1, x_2, x_3, \ldots$. Thus we want

$$y_n \approx y(x_n) \tag{3}$$

for each n.

When $x = x_0$, the rate of change of y with respect to x is $y' = f(x_0, y_0)$. If $y(x)$ continued to change at this same rate from $x = x_0$ to $x_1 = x_0 + h$, the change in the value of y would be $f(x_0, y_0)h$. We therefore take

$$y_1 = y_0 + f(x_0, y_0)h \tag{4}$$

as our approximation to the true value $y(x_1)$ of the solution at $x = x_1$. Similarly, we take

$$y_2 = y_1 + f(x_1, y_1)h \tag{5}$$

as our approximation to $y(x_2)$. Having reached the nth approximate value $y_n \approx y(x_n)$, we take

$$y_{n+1} = y_n + f(x_n, y_n)h \tag{6}$$

as our approximation to the true value $y(x_{n+1})$.

In summary, *Euler's method with step size h* for the initial value problem

$$\frac{dy}{dx} = f(x, y), \qquad y(x_0) = y_0 \tag{1}$$

consists in applying the iterative formula

$$y_{n+1} = y_n + hf(x_n, y_n) \tag{6}$$

to compute successive approximations $y_1, y_2, y_3, \ldots$ to the (true) values $y(x_1), y(x_2), y(x_3), \ldots$ of the (exact) solution $y = y(x)$ at the points $x_1, x_2, x_3, \ldots$ (where $x_{n+1} = x_n + h$).

Although the most important practical applications of numerical methods are to nonlinear differential equations, we first illustrate Euler's method with the linear initial value problem

$$\frac{dy}{dx} = x + y, \qquad y(0) = 1, \tag{7}$$

in order that we may compare our approximate solution with the exact solution

$$y(x) = 2e^x - x - 1, \tag{8}$$

which is easily found by standard methods (see Section 18.3 of Edwards and Penney, *Calculus*).

Example 1

Apply Euler's method to approximate the solution of the initial value problem in (7) on the interval $0 \leq x \leq 1$. Use successively the step sizes $h = 0.1, h = 0.02, h = 0.004$.

Solution   Here $f(x, y) = x + y$, so Euler's iterative formula is

$$y_{n+1} = y_n + h(x_n + y_n). \tag{9}$$

```
100 REM--Program EULER
110 REM--Uses Euler's method to approximate the solution
120 REM--of the equation y'=f(x,y) on the interval [0,1].
130 REM--The function f(x,y) is edited into line 230, and
140 REM--and the number N of subintervals is input.
150 REM
160 REM--Initialization:
170     PRINT "EDIT IN YOUR FUNCTION, THEN RUN 190"
180     PRINT   :   EDIT 230
190     PRINT "EDIT IN EXACT SOLUTION, THEN RUN 210"
200     PRINT   :   EDIT 240
210     DEFDBL A,F,G,H,X,Y   :   DEFINT I,J,K,N
220     DIM A(3,10)
230     DEF FNF(X,Y) = X + Y
240     DEF FNG(X)   = 2*EXP(X) - X - 1
250     INPUT "INITIAL VALUE Y(0)"; Y0
260     INPUT "INITIAL NUMBER OF SUBINTERVALS"; N
270     K = N/10  :  H = 1/N  :  X0 = 0
280     PRINT "     Y with     Y with     Y with"
290     PRINT " X    h=0.1      h=0.02     h=0.004    Y exact"
300     PRINT
310 REM
```

Listing 5.24   Program EULER

```
320 REM--Euler's iteration:
330       FOR I = 1 TO 3
340           X = XO   :   Y = YO
350           FOR J = 0 TO N
360               IF J/K = J\K THEN A(I,J/K) = Y
370               Y = Y + H*FNF(X,Y)
380               X = X + H
390           NEXT J
400           H = H/5  :  K = 5*K  :  N = 5*N
410       NEXT I
420 REM
430 REM--Print results:
440       FOR J = 0 TO 10
450           PRINT USING
              "#.#   #.#####   #.#####   #.#####   #.#####";
              J/10, A(1,J), A(2,J), A(3,J), FNG(J/10)
460       NEXT J
470       END
```

Listing 5.24 (con't.)

Program EULER (Listing 5.24) implements this iteration. Upon execution, the function F(X, Y) is edited into line 230, and the exact solution G(X) (if known) is edited into line 240. The values I = 1, 2, 3 in the outer FOR-NEXT loop (lines 330–410) correspond to the step sizes $h = 0.1$, $h = 0.02$, and $h = 0.004$, respectively. If the initial number N = 10 of subintervals is input at line 260, the inner FOR-NEXT loop (lines 350–390) applies Euler's method with N = 10, N = 50, and N = 200 steps (for I = 1, 2, 3, respectively) from $x = 0$ to $x = 1$. Note that Euler's iterative formula itself appears in line 370. When x is an integral multiple of 0.1, the corresponding approximate value of y is stored (with J = 10x) in the array A(I, J) for later printing (lines 440–460).

```
INITIAL VALUE Y(0)? 1
INITIAL NUMBER OF SUBINTERVALS? 10
          Y with     Y with     Y with
   X      h=0.1      h=0.02     h=0.004    Y exact

  0.0     1.00000    1.00000    1.00000    1.00000
  0.1     1.10000    1.10816    1.10990    1.11034
  0.2     1.22000    1.23799    1.24183    1.24281
  0.3     1.36200    1.39174    1.39810    1.39972
  0.4     1.52820    1.57189    1.58127    1.58365
  0.5     1.72102    1.78121    1.79416    1.79744
  0.6     1.94312    2.02272    2.03988    2.04424
  0.7     2.19743    2.29978    2.32189    2.32751
  0.8     2.48718    2.61608    2.64398    2.65108
  0.9     2.81590    2.97571    3.01038    3.01921
  1.0     3.18748    3.38318    3.42573    3.43656
```

Figure 5.25   Using Euler's method to approximate the solution of (7)

Figure 5.25 shows the output of Program EULER for the initial value problem in (7). Observe that for each step size h, the error (y actual − y approx) increases as x increases (that is, is farther from x_0). However, these errors decrease as h decreases. The percentage errors at the final point $x = 1$ are 7.25

percent with $h = 0.1$, 1.55 percent with $h = 0.02$, and only 0.32 percent with $h = 0.004$.

The Improved Euler Method

To increase the accuracy of the basic Euler technique, we can regard the result

$$u_{n+1} = y_n + hf(x_n, y_n) \tag{10}$$

of Euler's formula as merely a first attempt at estimating the true value of $y(x_{n+1})$. We then use the average value

$$m_n = \frac{1}{2}[f(x_n, y_n) + f(x_{n+1}, u_{n+1})] \tag{11}$$

as the rate of change of y with respect to x on the interval $[x_n, x_{n+1}]$. This yields

$$y_{n+1} = y_n + hm_n \tag{12}$$

as our $(n + 1)$st approximate value.

The *improved Euler method* consists of applying Formulas (10) through (13) at each step in the iteration to compute the successive approximations $y_1, y_2, y_3, \ldots$. It is an example of a *predictor-corrector method*. Formula (10) is the *predictor* used as a first attempt at the next approximation. Formula (12) is the *corrector* used to correct the value of the first attempt.

In order to convert Program EULER to an improved Euler method program, it is necessary only to replace line 370 with the two lines

```
370    U = Y + H*FNF(X, Y)
375    Y = Y + H*(0.5)*(FNF(X, Y) + FNF(X + H, U))
```

and to insert U in the list of double-precision variables declared in line 210. Figure 5.26 shows the improved Euler results for the initial value problem of Example 1. At the final point $x = 1$ the percentage errors now are 0.244 percent with $h = 0.1$ and 0.010 percent with $h = 0.02$, while with $h = 0.004$ the results shown are accurate to within a unit or two in the fifth decimal place—a rather impressive improvement.

We now apply the improved Euler method to an initial value problem that illustrates some of the possible pitfalls in numerical approximation.

Example 2

Approximate for $x > 0$ the solution of the initial value problem

$$\frac{dy}{dx} = x^2 + y^2, \qquad y(0) = 1. \tag{13}$$

```
INITIAL VALUE Y(O)? 1
INITIAL NUMBER OF SUBINTERVALS? 10
          Y with      Y with      Y with
   X      h=0.1       h=0.02      h=0.004     Y exact

  0.0    1.00000     1.00000     1.00000     1.00000
  0.1    1.11000     1.11033     1.11034     1.11034
  0.2    1.24205     1.24277     1.24280     1.24281
  0.3    1.39847     1.39966     1.39972     1.39972
  0.4    1.58180     1.58357     1.58365     1.58365
  0.5    1.79489     1.79733     1.79744     1.79744
  0.6    2.04086     2.04409     2.04423     2.04424
  0.7    2.32315     2.32732     2.32750     2.32751
  0.8    2.64558     2.65085     2.65107     2.65108
  0.9    3.01236     3.01892     3.01919     3.01921
  1.0    3.42816     3.43621     3.43655     3.43656
```

Figure 5.26   Using the improved Euler method to approximate the solution of (7)

```
INITIAL VALUE Y(O)? 1
INITIAL NUMBER OF SUBINTERVALS? 10
          Y with      Y with      Y with
   X      h=0.1       h=0.02      h=0.004     Y exact

  0.0    1.00000     1.00000     1.00000
  0.1    1.10000     1.10884     1.11092
  0.2    1.22200     1.24580     1.25152
  0.3    1.37533     1.42433     1.43648
  0.4    1.57348     1.66584     1.68975
  0.5    1.83707     2.00739     2.05429
  0.6    2.19955     2.52009     2.61695
  0.7    2.71935     3.36123     3.58651
  0.8    3.50783     4.96009     5.62735
  0.9    4.80232     8.99995     %12.55462
  1.0    7.18955     %30.91672   %40134.46025
```

Figure 5.27   Using the Euler method to approximate the solution of (13)

Solution   First we apply the original Euler method on $0 \leq x \leq 1$ by running Program EULER with $f(x, y) = x^2 + y^2$. The last line of data in the table of Figure 5.27 indicates that something exotic is happening near $x = 1$. Indeed, it turns out that $y(x) \longrightarrow +\infty$ as $x \longrightarrow 0.969811$ (from the left), as indicated in Figure 5.28.

Program IMPEULER (Listing 5.29) enables us to trace the behavior of the solution $y(x)$ as x approaches the singularity corresponding to the vertical asymptote in Figure 5.28. When an initial interval $[x_0, x_1]$ and an initial value $y_0 = y(x_0)$ are input, the main program (lines 260–310) calls on the subroutine in lines 330–450 to apply the improved Euler method to calculate $y_1 \approx y(x_1)$. This is done several times in succession, using at each stage five times the previous number of subintervals of $[x_0, x_1]$. The iteration is continued until two successive approximations (YNEW and YOLD) to y_1 agree to within the error tolerance that is checked in line 430. After the result is printed (line 300) we are offered the option (line 470) of continuing with a new interval $[x_0, x_1]$, in which case y_1 is used as the new initial value y_0. Thus we can apply the improved Euler method on successively shorter intervals as we approach the singularity.

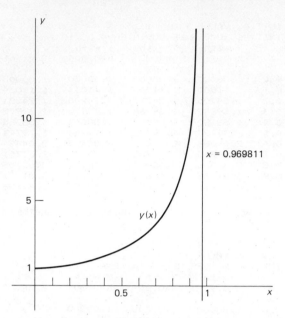

Figure 5.28   Graph of the solution $y(x)$ of the initial value problem (13)

```
100 REM--Program IMPEULER
110 REM--Uses the improved Euler method to
120 REM--approximate the solution of the
130 REM--differential equation y' = f(x,y) on
140 REM--the input interval. The function f
150 REM--is defined in line 190.
160 REM
170 REM--Initialization:
180      DEFDBL F,H,X,Y,U   :   DEFINT I,J,M,N
190      DEF FNF(X,Y) = X*X + Y*Y
200      LPRINT "          X                Y": LPRINT
210      INPUT "INITIAL VALUE Y0"; Y0
220      INPUT "ENDPOINTS X0,X1"; X0,X1
230      INPUT "PRINTING STEP SIZE"; HP
240      INPUT "ERROR TOLERANCE"; EPS
250 REM
260 REM--Main program:
270      M = (X1 -X0)/HP
280      FOR I = 1 TO M
290          GOSUB 330   'Improved Euler step
300          LPRINT USING
             "      #.####        #####.##"; X0,Y0
310      NEXT I   :   GOTO   470
320 REM
330 REM--Improved Euler subroutine:
340      H = HP    :    N = 1
350      YOLD = Y0 + H*FNF(X0,Y0)
360      X = X0    :    Y = Y0
370      FOR J = 1 TO N
380          U = Y + H*FNF(X,Y)
390          Y = Y + H*(.5)*(FNF(X,Y) + FNF(X+H,U))
400          X = X + H
```

Listing 5.29   Program IMPEULER

```
410        NEXT J
420        YNEW = Y    :    H = H/5    :    N = 5*N
430        IF ABS(YNEW - YOLD) > EPS*ABS(YNEW) THEN
               YOLD = YNEW   :     GOTO 360
440        XO = X   : YO = Y
450        RETURN
460 REM
470        INPUT "WANT TO CONTINUE (Y/N)"; Y$
480        IF Y$ = "Y" OR Y$ = "y" THEN GOTO 220
490        END
```

Listing 5.29 (con't.)

X	Y
0.1000	1.11
0.2000	1.25
0.3000	1.44
0.4000	1.70
0.5000	2.07
0.6000	2.64
0.7000	3.65
0.8000	5.85
0.9000	14.30
0.9100	16.70
0.9200	20.06
0.9300	25.10
0.9400	33.53
0.9500	50.46
0.9600	101.86
0.9610	113.41
0.9620	127.92
0.9630	146.69
0.9640	171.91
0.9650	207.59
0.9660	261.98
0.9670	354.97
0.9680	550.32
0.9690	1223.83
0.9691	1394.49
0.9692	1620.45
0.9693	1933.82
0.9694	2397.44
0.9695	3153.46
0.9696	4605.91
0.9697	8538.81
0.9698	58437.50

Figure 5.30 Using Program IMPEULER
to approximate the solution of (13)

Figure 5.30 shows the output of a run (with EPS = 0.0001) of Program IMPEULER to approximate the solution of the initial value problem of Example 2. Our rather unsophisticated program does not incorporate sufficient error controls to prevent substantial accumulation of errors in a problem like this one. Consequently, we cannot be confident that the value of $y(0.9698)$ is actually 58437.50, only that its magnitude is quite large. Nevertheless, the table provides strong empirical evidence that the solution $y(x)$ blows up near $x = 0.9698$. In the final step of the run of Program IMPEULER to produce these data, the interval $[0.9697, 0.9698]$ was subdivided into 3125 subintervals of length 3.2×10^{-8}. Figure 5.31 shows the input that produced the data shown in Figure 5.30.

```
RUN
INITIAL VALUE Y0? 1
ENDPOINTS X0,X1? 0, 0.9
PRINTING STEP SIZE? 0.1
ERROR TOLERANCE? 0.0001
WANT TO CONTINUE (Y/N)? Y
ENDPOINTS X0,X1? 0.9, 0.96
PRINTING STEP SIZE? 0.01
ERROR TOLERANCE? 0.0001
WANT TO CONTINUE (Y/N)? Y
ENDPOINTS X0,X1? 0.96, 0.969
PRINTING STEP SIZE? 0.001
ERROR TOLERANCE? 0.0001
WANT TO CONTINUE (Y/N)? Y
ENDPOINTS X0,X1? 0.969, 0.9698
PRINTING STEP SIZE? 0.0001
ERROR TOLERANCE? 0.0001
WANT TO CONTINUE (Y/N)? N
```

Figure 5.31 Input for run of Program IMPEULER

PROBLEMS

In each of Problems 1-10, use the improved Euler method (Program EULER with lines 370-375 inserted) to approximate the solution $y(x)$ given for the indicated initial value problem on the interval $[0, 1]$.

1. $y' = -y,$ $\qquad y(0) = 2;$ $\qquad y(x) = 2e^{-x}$

2. $y' = 2y,$ $\qquad y(0) = 0.5;$ $\qquad y(x) = 0.5e^{2x}$

3. $y' = y + 1,$ $\qquad y(0) = 1;$ $\qquad y(x) = 2e^x - 1$

4. $y' = x - y,$ $\qquad y(0) = 1;$ $\qquad y(x) = 2e^{-x} + x - 1$

5. $y' = y - x - 1,$ $\qquad y(0) = 1;$ $\qquad y(x) = 2 + x - e^{-x}$

6. $y' = -2xy,$ $\qquad y(0) = 2;$ $\qquad y(x) = 2\exp(-x^2)$

7. $y' = -3x^2y,$ $\qquad y(0) = 3;$ $\qquad y(x) = 3\exp(-x^3)$

8. $y' = e^{-y},$ $\qquad y(0) = 0;$ $\qquad y(x) = \ln(x + 1)$

9. $y' = \dfrac{1 + y^2}{4},$ $\qquad y(0) = 1;$ $\qquad y(x) = \tan\dfrac{x + \pi}{4}$

10. $(1 + 3y^2)y' = 2x,$ $\qquad y(0) = 1;$

   Suggestion: $y(x)$ satisfies the equation $y^3 + y = x^2 + 2$. Use a Newton's method subroutine to solve for $y(x)$.

11. Use Program IMPEULER as in Example 2 to verify empirically that the solution of the initial value problem

$$y' = x^2 + y^2, \qquad y(0) = 0$$

has a vertical asymptote near $x = 2.003147$.

12. The general solution of the equation

$$y' = (1 + y^2) \cos x$$

is $y(x) = \tan(C + \sin x)$. With the initial condition $y(0) = 0$ the solution $y(0) = \tan(\sin x)$ is well behaved. But with $y(0) = 1$ the solution $y(x) = \tan(\pi/4 + \sin x)$ has a vertical asymptote at $x = \sin^{-1}(\pi/4) \approx 0.90334$. Use Program IMPEULER to verify this fact empirically.

Further Reading

For a standard calculus reference I can recommend no better choice than

C. H. Edwards, Jr., and D. E. Penney, *Calculus and Analytic Geometry*, second edition, Prentice-Hall, Englewood Cliffs, N.J., 1986.

An excellent BASIC programming text is

T. Dwyer and M. Critchfield, *BASIC and the Personal Computer*, Addison-Wesley, Reading, Mass., 1978.

A very readable book on both elementary BASIC and the IBM Personal Computer is

L. G. Goldstein and M. Goldstein, *IBM PC: An Introduction to the Operating System, BASIC Programming and Applications*, Robert J. Brady Co., Bowie, Md., 1984.

For an especially wide range of applications of calculus and BASIC, see the book

W. R. Bennett, Jr., *Scientific and Engineering Problem-Solving with the Computer*, Prentice-Hall, Englewood Cliffs, N.J., 1976.

to which I referred in Chapter 5. Many readers who finish *Calculus and the Personal Computer* will want to move on from BASIC to a more powerful programming language. FORTRAN has served for some time as the standard general-purpose language for scientific computing on mainframes, but does not enjoy the same favor among microcomputer users. To learn a good deal of both Pascal and mathematics at the same time, I recommend

H. Flanders, *Scientific Pascal*, Reston, Reston, Va., 1984.

However, Pascal is a compiled language (though the Turbo Pascal recently introduced by Borland International, Inc. is a vast improvement over previous versions), and I feel that an interpreted language is needed to exploit fully the power and immediacy of the personal computer for interactive problem-solving. My own preference beyond BASIC is APL, a language that is designed specifically for the expression of mathematical algorithms. The standard APL programming text is

L. GILMAN and A. J. ROSE, *APL, An Interactive Approach*, third edition, John Wiley, New York, 1984.

Although there is not yet (to my knowledge) a book on APL and calculus, a fascinating blend of APL and mathematics is found in

G. HELZER, *Applied Linear Algebra with APL*, Little, Brown, Boston, 1983.

Program Listings

Computerized Practice Testing of Differentiation and Integration Techniques

Appendix

1

Here we present a simple system for student use of a personal computer to practice formal differentiation and integration skills. This system can be used by an individual student with access to a personal computer, or can serve as a starting point for a calculus instructor or computer laboratory coordinator who wishes to develop a more complete and permanent testing system.

When a student runs Program PRACTEST (listed below) and enters his name, ID number, and the current time, he first is asked whether he wishes to practice differentiation or integration (D or I?). If he selects D he then is presented with the menu

Level 1:   Polynomials and Rational Functions
Level 2:   Algebraic Functions
Level 3:   Exponential and Logarithmic Functions
Level 4:   Trigonometric and Inverse Trig Functions

If he selects I he is presented with the menu

Level 1:   Elementary Integrals and Substitutions
Level 2:   Trigonometric Integrals
Level 3:   Integration by Parts
Level 4:   Trigonometric Substitution
Level 5:   Integrals Involving Quadratic Functions
Level 6:   Rational Functions and Partial Fractions

Once the student has entered his choice, he is presented with a practice test consisting of ten randomly selected problems of the selected type and level.

```
What do you wish to practice?
Select either  (D)ifferentiation
          or   (I)ntegration

D or I? D
Select the level of differentiation skills that you wish to test:

              Level 1:  Polynomials and Rational Functions
              Level 2:  Algebraic Functions
              Level 3:  Exponential and Logarithmic Functions
              Level 4:  Trigonometric and Inverse Trig Functions

Desired Level  (1 to 4)? 1
LEVEL 1  DIFFERENTIATION TEST

This test consists of 10 problems.
You are allowed 2 attempt(s) on each problem.

Press any key when you are ready to begin.
Problem 1:

Suppose the formula for the function F(X) is

1/X

Then type (in BASIC notation) and enter a formula for
the derivative F'(X) of the function F(X):

Derivative? 1/(X*)X

Your answer cannot be evaluated.
Please reenter the formula for the derivative F'(X):
Derivative? 1/(X*X)

Wrong, try again!
Please reenter the formula for the derivative F'(X):
Derivative? -1/(X*X)

You got Problem 1 right.

SCORE:     1   RIGHT
           0   WRONG
           2   ATTEMPTS

Press any key when ready for next problem.
```

Figure A.1 Typical PRACTEST screen displays

Figure A.1 illustrates the screen displays that appear as the student proceeds through the test. Here it is assumed that the student is allowed two attempts per problem. The instructor determines in advance the allowed number of attempts by selecting the value of M in line 180 of PRACTEST; if $M = 1$ then only one attempt per problem is allowed.

Program PRACTEST is somewhat longer and more complicated than the other programs in this book and hence we only describe here several of its principal features. The program uses ten random access files corresponding to the ten available types and levels of test: DERIV1, . . . , DERIV4, INTGRL1, . . . , INTGRL6. Each record in each file consists of four character strings: A$ of length 4, B$ of length 4, F$ of length 60, and H$ of length 60. The string F$ is the formula for the function F(X) to be differentiated (or integrated), and H$ is the formula for its derivative (or integral). The strings A$ and B$ are the character representations of single-precision numbers A and B, where [A, B] is an interval on which both the function F(X) and its derivative (or integral) are defined. In lines 400–450 the records in the appropriate file (opened for random access as #1) are randomly ordered. The first N = 10 items in this order comprise the test that is administered in the main loop consisting of lines 590–1080.

The Ith problem (I = 1, 2, . . . , N) for the student to solve is displayed on the screen by lines 660–680, and the student inputs his answer—the formula G$ for the derivative (or integral) of the function F(X)—at line 760 or 770. The student's answer G(X) and the correct answer H(X) must be defined in lines 850–860. This is the purpose of lines 780–840, where the new versions of lines 850–860 are first stored in a temporary sequential file (opened as #2), and then chain merged into the main program. (This procedure results in considerable disk access during execution of Program PRACTEST; use of a RAM disk would speed things up.)

The student's answer G(X) and the correct answer H(X) + C (where C = 0 in the case of differentiation; otherwise, C is the appropriate constant of integration) are compared in the subroutine consisting of lines 1240–1320. If the two answers agree with relative error at most EPS = 0.0001 (specified in line 190) at each of 10 randomly selected points of the interval [A, B], then the student's answer is scored as "right" (F = 1); otherwise, it's "wrong" (F = 0).

Although Program PRACTEST may not be totally bombproof, the error-trapping routine in lines 1700–1840 will catch most things that can go wrong. For instance, if the student inadvertently enters a string of characters that doesn't even define a function at all (let alone the right one), then the ON ERROR GOTO 1700 command leads ultimately to lines 1780–1810, where he is asked to try again.

The program maintains a running total of right answers, wrong answers, and attempts. At the conclusion of the test a score sheet is printed, listing these scores under the type and level of the test, the student's name and ID, the date and time, and the time elapsed.

For systematic use in a computer laboratory setting, various refinements to Program PRACTEST might be useful. For example, the test problems and the student's answers could be stored in arrays as the test proceeds, and then at the end a copy of these problems and answers printed for the student or instructor to review.

Program ADDFCTNS (Listing A.3) is used to construct and maintain the "test bank" files that are accessed by Program PRACTEST. When ADDFCTNS is loaded and run, it first asks for the type and level of the file to

```
100 'Program PRACTEST
110 'For practice and testing of differentiation
120 'and integration skills.
130 '.
140 'Initialization:
150 ON ERROR GOTO 1700
160 DEFINT F,I,J,K,L,M,N,T
170 N = 10    :    'Number of problems per test
180 M = 2     :    'Number of attempts per problem
190 EPS = .0001  :  'Error tolerance for comparisons
200 CLS    :    KEY OFF
210 PRINT "To take a practice test, first enter your"
220 PRINT "name, ID number, and the current time."
230 INPUT "Your name"; NME$
240 INPUT "Your ID number"; ID$
250 INPUT "Time in hh:mm format"; TIMEX$
260 TIME$ = TIMEX$
270 PRINT : PRINT "What do you wish to practice?"
280 PRINT "Select either (D)ifferentiation"
290 PRINT "          or (I)ntegration"
300 PRINT  :  INPUT "D or I"; T$
310 IF T$ = "D" OR T$ = "d" THEN T = 1 ELSE
    IF T$ = "I" OR T$ = "i" THEN T = 2 ELSE
        PRINT : PRINT "Try again!" : GOTO 300
320 IF T = 1 THEN GOSUB 1340  :  'To select level L and
330 IF T = 2 THEN GOSUB 1500  :  'open functions file #1
340 FIELD #1, 4 AS A$, 4 AS B$, 60 AS F$, 60 AS H$
350 '(F = function, H = derivative or integral on [A,B])
360 NFCTNS = LOF(1)/128  :  'No of functions in file
370 DIM INDEX(NFCTNS)
380 RANDOMIZE TIMER
390 '
400 'To randomly order functions in file:
410 FOR I=1 TO NFCTNS : INDEX(I) = I : NEXT I
420 FOR I = NFCTNS TO 2 STEP -1
430     J = 1 + INT(RND*NFCTNS)
440     SWAP INDEX(I), INDEX(J)
450 NEXT I
460 '
470 CLS
480 IF T = 1 THEN PRINT "LEVEL";L;" DIFFERENTIATION TEST"
490 IF T = 2 THEN PRINT "LEVEL";L;" INTEGRATION TEST"
500 PRINT
510 PRINT "This test consists of";N;" problems."
520 PRINT "You are allowed";M;"attempt(s) on each problem."
530 PRINT
540 PRINT "Press any key when you are ready to begin."
550 IF INKEY$ = "" THEN GOTO 550
560 START = TIMER
570 NRITE = 0  :  NRONG = 0  :  NATTMPT = 0
580 '
```

Listing A.2   Program PRACTEST

```
590  'Main loop:
600  '
610   I = 1
620  'FOR I = 1 TO N
630    CLS
640    GET #1, INDEX(I)
650    A = CVS(A$)   :   B = CVS(B$)   :   'Endpoints
660    PRINT "Problem";I;":"   :   PRINT
670    PRINT "Suppose the formula for the function F(X) is"
680    PRINT    :   PRINT F$   :   PRINT
690    F = 0   :   K = 0      :   'F = 1 when answer correct
700   'WHILE F = 0 AND K < M
710      PRINT "Then type (in BASIC notation) and enter ";
720      PRINT "a formula for"
730      IF T = 1 THEN PRINT
           "the derivative F'(X) of the function F(X):"
740      IF T = 2 THEN PRINT
           "an indefinite integral of the function F(X):"
750      PRINT
760      IF T = 1 THEN INPUT "Derivative"; G$   :   PRINT
770      IF T = 2 THEN INPUT "Integral"; G$    :   PRINT
780      OPEN "FCTNS.BAS" FOR OUTPUT AS #2
790      L1$ = "850 DEF FNG(X) = "+G$+CHR$(13)+CHR$(10)
800      L2$ = "860 DEF FNH(X) = "+H$+CHR$(13)+CHR$(10)
810      PRINT #2, L1$   :    PRINT #2, L2$
820      CLOSE #2
830      CHAIN MERGE "FCTNS.BAS", 850, ALL
840      KILL "FCTNS.BAS"
850      DEF FNG(X) =
860      DEF FNH(X) =
870      ON ERROR GOTO 1700
880      C = 0   :   IF T = 1 THEN GOTO 910
890      X0 = (A + B)/2
900      C = FNG(X0) - FNH(X0)   :   'Constant of integration
910      GOSUB 1240   :   'To compare G(X) and H(X)
920      IF F = 0 AND K < M-1 THEN PRINT "Wrong, try again!"
930      K = K+1   :    NATTMPT = NATTMPT + 1
940      IF F = 0 AND K < M THEN GOTO 710
950   'WEND
960    IF F = 1 THEN NRITE = NRITE + 1   :   PRINT
         "You got Problem";I;" right."   :   PRINT   :
         GOTO 1000
970    IF F = 0 THEN NRONG = NRONG + 1   :   PRINT
         "You got Problem";I;" wrong."
980    PRINT "A correct answer is"              :   PRINT
990    PRINT H$     :    PRINT
1000   PRINT "SCORE:      "; NRITE; " RIGHT"
1010   PRINT "           "; NRONG; " WRONG"
1020   PRINT "           "; NATTMPT; " ATTEMPTS"
1030   PRINT
```

Listing A.2 (con't.)

```
1040    PRINT "Press any key when ready for next problem."
1050     IF INKEY$ = "" THEN GOTO 1050
1060      I = I + 1    :  IF I <= N THEN GOTO 630
1070   'NEXT I
1080   'Endloop
1090   '
1095   'Print test score:
1100    CLS
1110    PRINT NME$    :    PRINT ID$
1115    PRINT "Date:  ";DATE$,"Time:   ";TIME$
1120    ELAPSED = TIMER - START
1130    PRINT (ELAPSED\60); " minutes elapsed"
1140    IF T = 1 THEN PRINT "Level";L;"Differentiation Test"
1150    IF T = 2 THEN PRINT "Level";L;"Integration Test"
1160    PRINT ,NRITE; "problems right"
1170    PRINT ,NRONG; "problems wrong"
1180    PRINT ,NATTMPT; "total attempts"
1190   '
1200   INPUT "Want to take another test"; Y$
1210   IF Y$ = "Y" OR Y$ = "y" THEN GOTO 200
1220   GOTO 1860
1230   '
1240   'Comparison subroutine:  Returns F = 1 if  H(X) + C
1250   'and G(X) agree, F = 0 otherwise.
1260   F = 1
1270   FOR J = 1 TO 10
1280       X = A + RND*(B - A)
1290       IF ABS(FNH(X)) < EPS THEN GOTO 1280
1300       IF ABS((FNG(X) - FNH(X) - C)/(FNH(X)+C)) > EPS
              THEN F = 0   :   GOTO 1320
1310   NEXT J
1320   RETURN
1330   '
1340   'Select level of differentiation test:
1350   PRINT "Select the level of differentiation ";
1360   PRINT "skills that you wish to test:"   :   PRINT
1370   PRINT ,"Level 1:   Polynomials and Rational Functions"
1380   PRINT ,"Level 2:   Algebraic Functions"
1390   PRINT ,"Level 3:   Exponential and Logarithmic Functions"
1400   PRINT ,"Level 4:   Trigonometric and Inverse Trig Functions"
1410   PRINT
1420   INPUT "Desired level (1 to 4)"; L
1430   IF (L<>1 AND L <>2) AND (L<>3 AND L<>4) THEN
              PRINT "Try again!"   :   GOTO 1420
1440   IF L = 1 THEN OPEN "DERIV1" AS #1
1450   IF L = 2 THEN OPEN "DERIV2" AS #1
1460   IF L = 3 THEN OPEN "DERIV3" AS #1
1470   IF L = 4 THEN OPEN "DERIV4" AS #1
1480   RETURN
1490   '
```

Listing A.2 (con't.)

```
1500 'Select level of integration test:
1510 PRINT "Select the level of integration";
1520 PRINT "skills that you wish to test:"  :    PRINT
1530 PRINT ,"Level 1:  Elementary Integrals and Substitutions"
1540 PRINT ,"Level 2:  Trigonometric Integrals"
1550 PRINT ,"Level 3:  Integration by Parts"
1560 PRINT ,"Level 4:  Trigonometric Substitution"
1570 PRINT ,"Level 5:  Integrals Involving Quadratic Functions"
1580 PRINT ,"Level 6:  Rational Functions and Partial Fractions"
1590 PRINT
1600 INPUT "Desired level (1 to 6)"; L
1610 IF ((L<>1 AND L<>2) AND (L<>3 AND L<>4))
         AND (L<>5 AND L<>6) THEN
              PRINT "Try again!"  :   GOTO 1600
1620 IF L = 1 THEN OPEN "INTGRL1" AS #1
1630 IF L = 2 THEN OPEN "INTGRL2" AS #1
1640 IF L = 3 THEN OPEN "INTGRL3" AS #1
1650 IF L = 4 THEN OPEN "INTGRL4" AS #1
1660 IF L = 5 THEN OPEN "INTGRL5" AS #1
1670 IF L = 6 THEN OPEN "INTGRL6" AS #1
1680 RETURN
1690 '
1700 'Error trapping routine:
1710 IF ERR = 53 OR ERR = 71 THEN E = 1 ELSE
         IF ERL = 900 OR ERL = 1300 THEN E = 2 ELSE E = 3
1720 ON E GOTO 1730, 1780, 1820
1730 PRINT "Insert disk containing functions ";
1740 PRINT "files in Drive A."
1750 PRINT "Press any key when ready."
1760 IF INKEY$ = "" THEN 1760
1770 RESUME
1780 PRINT "Your answer cannot be evaluated."
1790 PRINT "Please reenter the formula for the ";
1800 IF T = T THEN PRINT "derivative F'(X)." :   RESUME 760
1810 IF T = 2 THEN PRINT "integral of F(X)." :   RESUME 770
1820 PRINT "Inappropriate input.  Please try again."
1830 IF ERL = 260 THEN RESUME 250
1840 RESUME
1850 '
1860 KEY ON
1870 END
```

Listing A.2 (con't.)

which you wish to add a new problem. When this selection has been entered, the program then prompts you to enter the new function F(X), its (correct) derivative or integral H(X), and the endpoints of an interval [A, B] on which both are defined. The program also provides the option of printing a copy of the current contents of a file. Before Program PRACTEST is run, at least ten function pairs must be stored in each of the test bank files that is to be used. On the optional diskette that is available (to instructors) with this book, files DERIV1, . . . , DERIV4, INTGRL1, . . . , INTGRL6 are loaded with routine differentiation and integration problems from Edwards and Penney, *Calculus.*

```
100  'Program ADDFCTNS
110  'To write and maintain the random access files that
120  'are used by Program PRACTEST.  These files are:
130  '
140  '            DERIV1, ------ ,DERIV4
150  '            INTGRL1, ------ ,INTGRL6
160  '
170   DEFINT I,O,L,N,T
180   CLS
190   PRINT "Insert the appropriate file diskette ";
200   PRINT "in Drive A."
210   PRINT "Press any key when ready to continue."
220   PRINT
230   IF INKEY$ = "" THEN GOTO 230
240   PRINT "Type of file: (D)erivatives or (I)ntegrals?"
250   INPUT "Type (D or I)"; T$
260   IF T$ = "D" OR T$ = "d" THEN T = 1 ELSE
      IF T$ = "I" OR T$ = "i" THEN T = 2 ELSE
         PRINT "Try again!"  :  GOTO 240
270   PRINT
280   PRINT "Enter level of file -- 1 to 4 for derivatives,"
290   PRINT "                       1 to 6 for integrals."
300   PRINT  :  INPUT "Level"; L  :  TYPE = T
310   ON TYPE GOSUB 810, 880  :  'Open approp file as #1
320   FIELD #1, 4 AS A$, 4 AS B$, 60 AS F$, 60 AS H$
330   '
340  'Options
350   CLS
360   PRINT "Your options are:"
370   PRINT
380   PRINT "    1 -- Add a function to the file
390   PRINT "    2 -- Print the file
400   PRINT "    3 -- Quit
410   PRINT
420   INPUT "Desired option (1 or 2 or 3)"; OPT
430   PRINT
440   ON OPT GOTO 460, 630, 970
450   '
460  'Add a function to file
470   N = 1 + LOF(1)/128
480   PRINT "Enter formula for new function F(X):"
490   PRINT  :  INPUT "Function"; FCTN1$  :  PRINT
500   PRINT "Enter formula for its derivative (or integral):"
510   PRINT
520   IF T = 1 THEN INPUT "Derivative"; FCTN2$
530   IF T = 2 THEN INPUT "Integral";   FCTN2$
540   PRINT
```

Listing A.3   Program ADDFCTNS

```
550    PRINT "Now enter endpoints of an interval [A,B]"
560    PRINT "on which both are defined:"
570    PRINT   :   INPUT "A,B"; A,B
580    LSET A$ = MKS$(A)   :   LSET B$ = MKS$(B)
590    LSET F$ = FCTN1$    :   LSET H$ = FCTN2$
600    PUT #1, N
610    GOTO 340
620    '
630    'Print the file:
640    CLS
650    PRINT "FILE:  TYPE = ";T$;"   LEVEL =";L
660    PRINT
670    N = LOF(1)/128   :   'Number of entries in file
680    FOR I = 1 TO N
690        PRINT "ENTRY"; I
700        GET #1, I
710        PRINT "A ="; CVS(A$), "B ="; CVS(B$)
720        PRINT "Function = ";F$
730        IF T = 1 THEN PRINT "Derivative = ";H$
740        IF T = 2 THEN PRINT "Integral = ";H$
750        PRINT
760    NEXT I
770    PRINT "Press any key when ready to continue."
780    IF INKEY$ = "" THEN GOTO 780
790    GOTO 340
800    '
810    'Select file of derivatives:
820    IF L=1 THEN OPEN "DERIV1" AS #1 LEN = 128
830    IF L=2 THEN OPEN "DERIV2" AS #1 LEN = 128
840    IF L=3 THEN OPEN "DERIV3" AS #1 LEN = 128
850    IF L=3 THEN OPEN "DERIV4" AS #1 LEN = 128
860    RETURN
870    '
880    'Select file of integrals:
890    IF L=1 THEN OPEN "INTGRL1" AS #1 LEN = 128
900    IF L=2 THEN OPEN "INTGRL2" AS #1 LEN = 128
910    IF L=3 THEN OPEN "INTGRL3" AS #1 LEN = 128
920    IF L=4 THEN OPEN "INTGRL4" AS #1 LEN = 128
930    IF L=5 THEN OPEN "INTGRL5" AS #1 LEN = 128
940    IF L=6 THEN OPEN "INTGRL6" AS #1 LEN = 128
950    RETURN
960    '
970    CLOSE #1
980    END
```

Listing A.3 (con't.)

Index